Building and Using Baluns and Ununs

Practical Designs For The Experimenter

Building and Using Baluns and Ununs
Practical Designs For The Experimenter

By Jerry Sevick, W2FMI

CQ Communications, Inc.

Third Printing 1996

Library of Congress Catalog Card Number 94-69520
ISBN 0-943016-09-6

Editor: Terry Littlefield, KA1STC
Managing Editor: Gail M. Schieber
Editorial Assistant: Nancy Barry
Layout and Design: Elizabeth Ryan
Illustrations: Hal Keith, K & S Graphics
Cover photo: Robert S. LeBlanc

Published by CQ Communications, Inc.
76 North Broadway
Hicksville, New York 11801 USA

Printed in the United States of America.

Preface

The transmission line transformer, which first appeared after World War II, is not considered (by the professionals) an emerging technology. Obviously, it has been in existence for a very long time. Although this technology has been widely used in solid-state circuitry and in matching transmission lines to antennas, there has been little information available that deals with the design and application of practical hardware. Many companies are reluctant to publish their results for fear of giving away their hard-earned secrets. Furthermore, this technology is not adequately covered in today's college curricula. Textbooks only contain a few paragraphs on the subject. Therefore, very few people fully understand this technology, and as a result, it is still far from reaching its full potential.

In an attempt to fill the need in this area, I published my first book[1], entitled *Transmission Line Transformers.** It not only reviewed the available theory on these devices, but also presented characterizations from extremely accurate data with fractional-ratio ununs (unbalanced-to-unbalanced transformers with ratios other than $1{:}n^2$ where n is 1, 2, 3, etc.), which were important to me in my short vertical antenna studies. Although the book was quite well received, it became very apparent that more specific information on practical designs was needed.

The feedback from the first edition, which I appreciated, led to the second edition,[2] which attempted to respond to the many requests for more practical design information. Therefore, some 100 new transformers were included in the second edition. In the process of crafting these new transformers, many interesting and unexpected designs emerged. This was especially true when using Guanella's approach of connecting transmission lines in series-parallel arrangements. (Guanella's method was published in his classic 1944 paper.[3]) However, the second edition did not really solve the problem for those who wanted to construct their own transformers, because many of the ferrite cores were not readily available. Although most of the ferrite manufacturers were more than willing to supply me with their samples for use in my designs, they were not in the retail business. In other words, they were reluctant to sell their cores in very small lots. By joining forces with Amidon Associates, Inc., I was able to redesign practically all of the transformers in the second edition with components that became readily available. Additionally, by careful standardization and redesign, I was not only able to improve the performances of many of the designs, but was also able to optimize the designs on a cost/performance basis.

Then, in recent articles in the amateur radio literature,[4-7] new terms appeared for the 1:1 and 4:1 baluns—*current, voltage,* and *choke* baluns. New balun designs using coaxial cables wound around a ferrite core or threaded through ferrite beads also appeared in these articles. In comparing these "new" baluns with the so-called transformer-type (wire transmission lines wound around a core) or voltage-type baluns (described later), the authors advanced various claims about which I had serious questions. As I looked further into the amateur radio literature, especially our handbooks, I found other points of disagreement.

As a result of reading this literature, I decided that a series of articles presenting my latest designs, and my views on 1:1 and 4:1 baluns in particular, would be of interest at this time. Fortunately, CQ Communications, Inc. provided the space in two of its journals,

**References through Chapter 11 are found listed in Chapter 12.*

resulting in an in-depth series on baluns in *Communications Quarterly* and a practical series on ununs and baluns in *CQ*. This book is essentially a combination of both series. It is also a complement to the second edition of *Transmission Line Transformers*.[2]

This book also contains in the appendixes some of my earlier work on ground systems and short verticals. Because this material was published over 20 years ago in *QST* and the antenna systems that followed utilized my unun design, I thought it appropriate to include it in this book. The fourth and last appendix describes my latest study on the ubiquitous loading coil. I am quite sure this appendix will be of interest to those engaged in mobile operation.

I wish to thank three people who were very instrumental in making this book possible. They are Alan M. Dorhoffer, K2EEK, Editor, *CQ* magazine; Terry Littlefield, KA1STC, Editor, *Communications Quarterly*; and James Lau, head of Amidon Associates, Inc. A second thanks is also extended to Terry Littlefield for her excellent technical help and cooperation in bringing this book to press.

Finally, this book is dedicated to my wife, Connie. She has given me the much needed support and time to complete this undertaking.

Jerry Sevick, W2FMI
Basking Ridge, New Jersey
December 1994

About The Author

Jerry Sevick, W2FMI, is well known in the amateur radio community for his work with transmission line transformers—particularly baluns and ununs. He is the author of the popular book *Transmission Line Transformers* published by The American Radio Relay League, a series on baluns in *Communications Quarterly*, and a series on ununs in *CQ*. His work with these devices has also been adopted as a standard by the IEEE.

Along with his work on transmission line transformers, Jerry is known for his experiments with short vertical antennas and broadband matching networks. He is noted for a classic series on short vertical antennas that appeared in *QST*, and his April 1978 *QST* article on short ground-radial systems serves as the world's standard for earth conductivity measurements.

Sevick holds a BS in education from Wayne State University and a PhD in applied physics from Harvard University. He taught physics at Wayne State from 1952 to 1956. In 1956, Jerry joined the staff at AT&T Bell Laboratories, where he supervised groups working in high-frequency transistor and integrated circuit development, reliability, applications engineering, and high-speed PCM. Jerry later served as Director of Technical Relations for AT&T Bell Laboratories until his retirement in 1985.

Dr. Sevick remains active in his field. He is currently a technical advisor for The American Radio Relay League and a member of the IEEE, Sigma Xi, Sigma Pi Sigma, and Phi Delta Kappa.

Table of Contents

Chapter 1

The 1:1 Balun

Sec 1.1 Introduction

In this first chapter of *Building Baluns and Ununs*, I will introduce the most popular broadband balun in amateur radio use—the 1:1 balun. This topic has been discussed in the amateur radio literature since the publication of Turrin's 1964 article.[8] Although Turrin's balun is really a version of Ruthroff's, which was introduced in his classic 1959 article,[9] the real beginning of the broadband 1:1 balun dates back to Guanella's classic paper of 1944.[3] Guanella's objective was to design a broadband 16:1 balun to match the balanced output impedance of 960 ohms of a push-pull, 100-watt vacuum-tube amplifier to the unbalanced load of a matched 60-ohm coaxial cable. Use of his approach for 1:1, 4:1, and 9:1 baluns has produced the designs of choice. They are presently called current or choke baluns.

This chapter begins with an introduction to the technology of transmission line transformers. Other topics discussed include: 1) when to use a balun; 2) highlights of significant articles in the professional and amateur radio literature; 3) high-power, medium-power, and low-power designs; and 4) isolation transformers. The latter is presented here for the first time. The chapter closes with a brief summary of the significant points included within. As you will see, the information available to radio amateurs has been sorely lacking over the past 25 years (at least)!

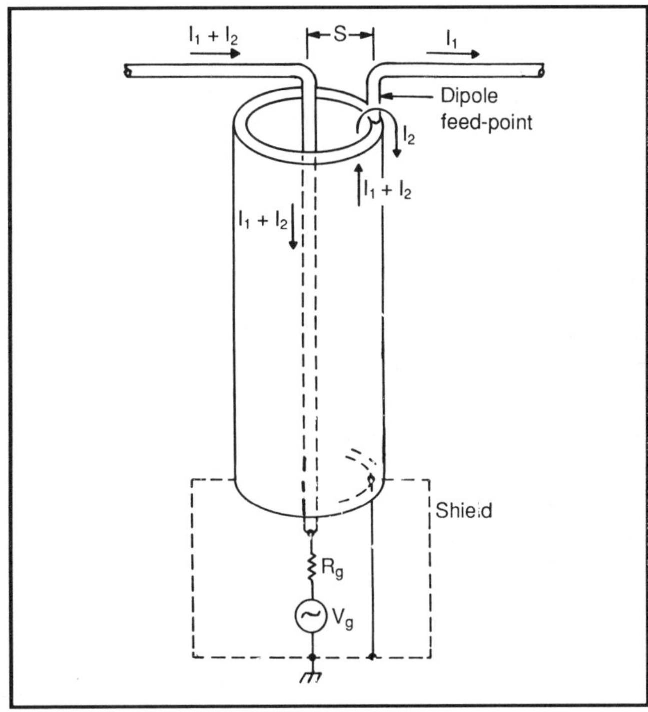

Figure 1-1. An illustration of the various currents at the feedpoint of a dipole. I_1 is the dipole current and I_2, the inverted L (imbalance) current.

Sec 1.2 When to Use a Balun

Baluns have taken on a more significant role in the past few decades with the advent of solid-state transceivers and Class B linear amplifiers with unbalanced outputs. That is, the voltage on the center conductor of their output chassis connectors varies (plus and minus) with respect to ground. In many cases, coaxial cables are used as the transmission lines from these unbalanced outputs to antennas like dipoles, inverted Vs, and Yagi beams that favor a balanced feed. In essence, they prefer a source of power whose terminals are balanced (voltages being equal and opposite) with respect to actual ground or to the virtual ground that bisects the center of the antenna. The question that is asked most frequently is whether a 1:1 balun is really needed.

To illustrate the problem involved and to give a basis for my suggestions, I refer you to **Figure 1-1**. Here we have, at the feedpoint of the dipole, two equal and opposite transmission line currents with two components each—I_1 and I_2. Also shown is the spacing, s, between the center conductor and the outside braid. Theoretically, a balanced antenna with a balanced feed would have a ground (zero potential) plane bisecting this spacing. However, because a coax-feed is unbalanced and the outer braid is also connected to ground at some point, an imbalance exists at the feedpoint giving rise to two antenna modes. One lies with I_1, providing a dipole mode; the other lies with I_2, providing an inverted L mode.

If the spacing, s, is increased, the imbalance at the feedpoint becomes greater—giving rise to more current on the outer braid and a larger imbalance of currents on the antenna's arms. Several steps can be taken to eliminate or minimize the undesirable inverted L mode (that is, eliminate or minimize I_2). The obvious choice is to use a well-designed balun that not only provides a balanced feed, but also minimizes (by its choking reactance) I_2—if the coaxial cable does not lie in the ground plane that bisects the center of the dipole. The other step is to ground the coaxial cable at a quarter-wave (or odd-multiple thereof) from the feedpoint. This discourages the inverted L mode because any radiating element will want to see a high impedance at these lengths instead of the low impedance of a ground connection.

I conducted experiments with baluns on a 20-meter half-wave dipole at a height of 0.17 wavelengths, which gave a resonant impedance of 50 ohms. VSWR curves were compared under various conditions. When the coaxial cable was in the ground plane of the antenna (that is, perpendicular to the axis of the antenna), the VSWR curves were identical with or without a well-designed balun—no matter where the outer braid was grounded. A significant difference was noted only when the coaxial cable was out of the ground plane. When the cable dropped down at a 45 degree angle under the dipole, a large change in the VSWR took place. This meant that the inverted L mode was appreciable.

It should also be mentioned that the direction of I_2, the imbalance current, can depend upon the side on which the coaxial cable is out of the ground plane of the dipole. For example, if the cable comes down under the right side in **Figure 1-1** (that is, the angle between the horizontal arm and the coax is less than 90 degrees on the right side and more than 90 degrees on the left side), then the direction of I_2 can be reversed by the imbalance in the induced currents on the outside of the braid. By the same token, by having the coaxial cable coming down on the other side, the value of I_2 is only increased in magnitude.

However, feeding a Yagi beam without a well-designed 1:1 balun is a different matter. Because most Yagi designs use shunt-feeding (usually by hairpin matching networks) in order to raise the input impedance close to 50 ohms, the effective spacing (s) is greatly increased. Furthermore, the center of the driven element is actually grounded. Thus, connecting the outer braid (which is grounded at some point) to one of the input terminals, creates a large imbalance and a real need for a balun. An interesting solution, which would eliminate the matching network, is to use a step-down balun designed to match 50-ohm cable directly to the lower balanced-impedance of the driven element.[2]

In summary, if you concur with the theoretical model of **Figure 1-1**, my experiments performed on 20 meters, and the reports from radio amateurs using dipoles and inverted Vs without baluns, then it appears that 1:1 baluns are really needed for: a) Yagi beam antennas where severe pattern distortion can take place without one and b) dipoles and inverted Vs that have the coaxial cable feed lines out of the ground plane that bisects the antennas, or that are unbalanced by their proximity to manmade or natural structures. In general, the need for a balun is not so critical with dipoles and inverted Vs (especially on 40, 80, and 160 meters) because the diameter of the coaxial cable connector at the feedpoint is much smaller than the wavelength.

If my model, which assumes that a part of the problem when feeding balanced antennas with coaxial cable is related to the size of the spacing, s (shown in **Figure 1-1**), then the possibility exists for using ununs for matching into balanced antennas with impedances other than 50 ohms and with small values of s. For example, half-wave dipoles at a height of about a half-wave, quads, and center-fed 3/2-wave dipoles—which all have impedances close to 100 ohms—could very well be matched to 50-ohm cable by a 2:1 unun. As I will show, they are considerably easier to construct than 2:1 baluns. Furthermore, Genaille[10] has recently shown considerable success using ununs in this kind of application.

To wind up this section, I would like to comment on an article published by Eggers,[11] WA9NEW, concerning the use of a balun with a half-wave dipole. While at North Carolina State University, he conducted an experimental investigation of pattern distortion without a balun at 1.6 GHz in an RF anechoic chamber (which simulates "free space"). Briefly, his results showed that, with a balun (bazooka type), the antenna radiation pattern compared very favorably with the classic "figure-eight." Without the balun, the radiation pattern was severely distorted.

Even though the author expressed difficulty in obtaining accurate measurements at this very high frequency, I have a question regarding the validity of performing the experiment in the first place. From the photograph in the article, it appears that conventional coaxial cable and connectors were used in the experiment. If we assume an effective diameter of 0.375 inches for these components, then scaling up to 3.5 MHz (457.14 fold) results in a coaxial cable with a diameter of 14.28 feet! I am quite sure that the large spacing, s, of 7.14 feet would bring about a noticeable imbalance resulting in appreciable pattern distortion even at 3.5 MHz.

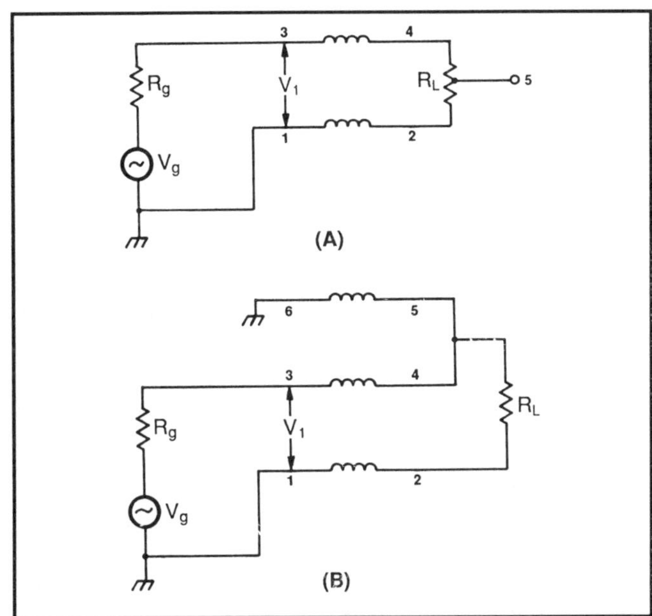

Figure 1-2. Two versions of the 1:1 balun: (A) The Guanella balun and the basic building block; (B) the Ruthroff balun as originally drawn.

Sec 1.3 Highlights of Significant Articles on 1:1 Baluns

Although there have been many articles on 1:1 baluns published in the professional and amateur literature, I have selected for review a few that I believe have had the most impact on 1:1 baluns for amateur radio use. As you will see, even though I consider some of the amateur articles significant, their impact upon the use and understanding of these devices has not always been positive. In fact, in some cases, the opposite has been true.

In the professional literature

There are actually only two significant articles in the professional literature that provide the fundamental principles upon which the theory and design of this class of transformers are based. It can be said that succeeding investigators simply extended the works of the authors of these two articles.

The first presentation on broadband matching transformers using transmission lines was given by Guanella in 1944.[3] He coiled transmission lines forming a choke such that only transmission line currents were allowed to flow, no matter where a ground was connected to the load. His single, coiled transmission line resulted in a 1:1 balun. It is shown schematically in **Figure 1-2A**. Prior to this, RF baluns were achieved by the use of quarter and half-wave transmission lines, and as a result, had narrow bandwidths. Guanella then demonstrated broadband baluns with impedance transformations of $1{:}n^2$ where n is the number of transmission lines he connected in a series-parallel arrangement.

Several important points should be made regarding Guanella's 1:1 balun shown in **Figure 1-2A**. With sufficient choking reactance, the output is isolated from the input and only flux-canceling transmission line currents are allowed to flow. With matched or very short transmission lines, the grounding of terminal 5 (actually or virtually like the center of a dipole) results in terminal 4 becoming $+V_1/2$ and terminal 2 becoming $-V_1/2$—creating a balanced output. This type of balun has lately been called a "current" or "choke" balun. A significant feature of this model is that a potential gradient of $-V_1/2$ exists along the length of the transmission line. This gradient, which exists on both conductors, accounts for practically all of the loss in these transformers because the loss mechanism is voltage dependent (a dielectric-type loss). All transmission line transformers have some sort of voltage gradient along their transmission lines and are, thus, subject to the same type of losses. Furthermore, the theory and loss mechanism is the same whether the transmission lines are coax or twin-lead, or coiled around cores, or threaded through fer-

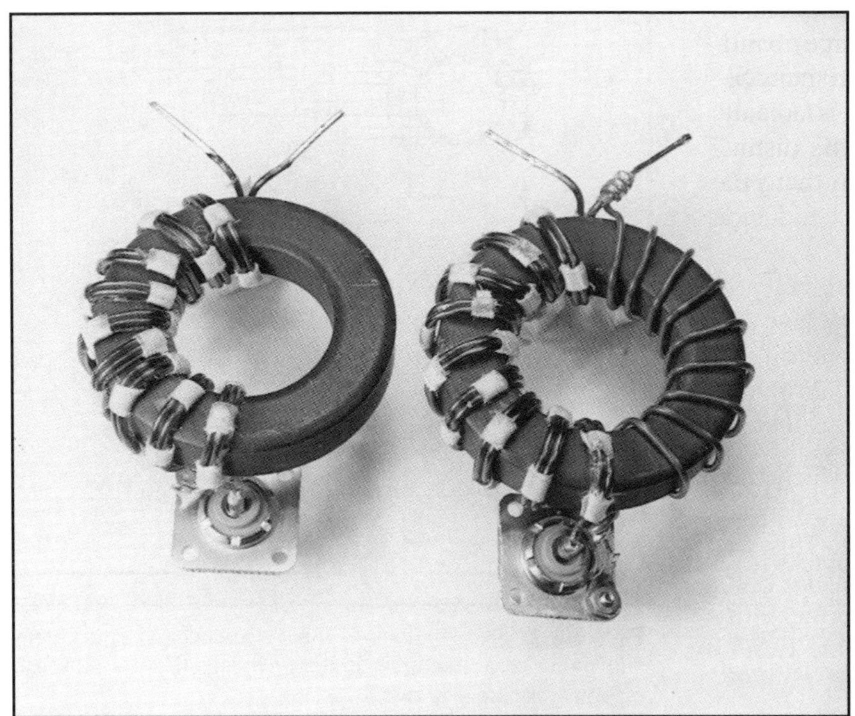

Photo 1-A. The two basic forms of the 1:1 balun that first appeared in the professional literature. The two-conductor Guanella balun is on the left and the three-conductor Ruthroff balun is on the right.

rite beads. Additionally, it was shown[2] that higher-impedance baluns or baluns subjected to higher VSWRs have more loss because the voltage gradients are also larger.

The second and other significant article on broadband transmission line transformers was published by Ruthroff.[9] His 1:1 balun, which is shown as originally drawn in **Figure 1-2B**, used an extra winding to complete (as he said) the path for the magnetizing current. Even though his schematic drawing appeared to look like a trifilar winding, his pictorial in the article clearly showed that the third winding (5-6) was on a separate part of the toroid. With an equal number of turns, it forms a voltage divider with winding (3-4) placing terminal 4 at $+V_1/2$ and terminal 2 at $-V_1/2$. In his classic paper, Ruthroff also presented his form of the 4:1 balun (which is also different from Guanella's), a 4:1 unun, and various hybrids. **Photo 1-A** shows the two basic forms of the 1:1 balun that first appeared in the professional literature. The two conductor Guanella 1:1 balun is on the left and the three conductor Ruthroff balun is on the right. As was mentioned before, the Guanella balun has recently been called a "current" or "choke" balun.

Before moving on to the significant articles in the amateur radio literature, some mention should be made of the differences between the two basic forms shown in **Photo 1-A**. Guanella's 1:1 balun came to be known as the *basic building block* for this whole class of broadband transformers. This term was coined by Ruthroff as he showed its 1:1 balun capability when the load was grounded at its center (terminal 5), and as a phase-inverter when the load was grounded at the top (terminal 4). By connecting terminal 2 to terminal 3 and connecting the bottom of the load to ground, Ruthroff then demonstrated his very popular 4:1 unun. I called this type of arrangement the "boot-strap" connection. By grounding terminal 2, there is no potential drop along the transmission line and therefore no need for magnetic cores or beads. This arrangement, which turns out to be an important function for extending the high frequency performance of this class of transformers, I call the "phase-delay" connection.

Thus, with the flexibility shown by Guanella's basic building block, a 1:1 balun is now realized that not only presents a balanced power source to a balanced antenna system, but can also prevent an imbalance current (an inverted L antenna current) by its choking reactance when the load is unbalanced or mismatched or when the feedline is not perpendicular to the axis of the antenna.

Interestingly enough, except at the very low end of the frequency response of the Ruthroff 1:1 balun where autotransformer action can take place, his balun takes on the characteristics of the Guanella balun. The reactance of the third winding becomes great enough to make it literally transparent. This is not the nature of the trifilar-wound (voltage) balun, which is sensitive to unbalanced and mismatched loads over its entire passband because it is actually two tightly coupled transmission lines. This distinction was not recognized by most of those that published in the amateur radio literature.

In the amateur radio literature

R. Turrin, W2IMU—1964

The first presentation in the amateur radio literature on 1:1 baluns using ferrite cores was by Turrin in 1964.[8] Turrin, who was a colleague of Ruthroff at Bell Labs, took his small-signal design (which used No. 37 or 38 wire on toroids with ODs of 0.25 inches or less) and adapted it to high-power use. This was done by using thicker wire, larger cores, and (very importantly for high efficiency[2]) low permeability ferrite. Ruthroff used lossy manganese-zinc ferrites with permeabilities of about 3000 because efficiency was not a major consideration.

Figure 1-3 shows a pictorial and a schematic of Turrin's design. As you can see, the third wire (winding 3-4) is placed between the two current-carrying wires (windings 1-2 and 4-5). **Photo 1-B** shows (on the left) his actual design using a ferrite core; a popular design (on the right) using a powdered-iron core is available in kit form from Amidon Associates, Inc. Both baluns use 10 trifilar turns of a single-coated wire such as Formex™ or Formvar™ on a toroid. Turrin's design uses a ferrite toroid with an OD of 2.4 inches and a permeability of 40. The Amidon Associates' balun uses a powdered-iron toroid with a 2-inch OD and a permeability of only 10. Both baluns

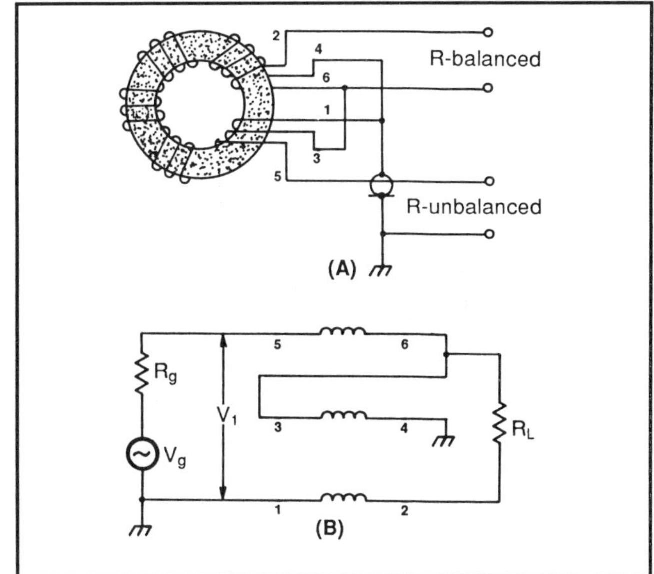

Figure 1-3. (A) A pictorial of Turrin's 1:1 balun, and (B) a schematic of his balun.

are specified to handle 1000 watts of power from 1.8 to 30 MHz.

Figure 1-4 shows the response curves for these two baluns when terminated with 50-ohm loads. The response curve for a popular 1:1 rod-type balun that uses the same schematic and wire is also shown. It has 8 trifilar turns, tightly wound on a rod of 0.5 inch diameter, 2.5 inches long, and with a permeability of 125. The rod-type balun is shown in **Photo 1-C**.

Several important features should be brought out regarding the results shown in **Figure 1-4**. They are:

1. All baluns had insufficient choking reactance and, hence, poor low-frequency responses. The powdered-iron version was especially poor. They all

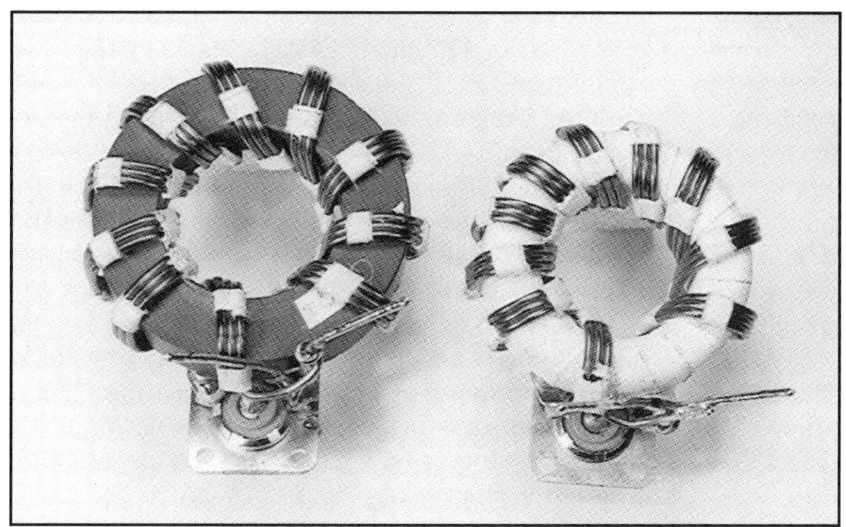

Photo 1-B. Two versions of Turrin's design: On the left, the 1:1 balun that has appeared in the amateur radio literature; on the right, a 1:1 balun that has been readily available in kit form from Amidon Associates, Inc.

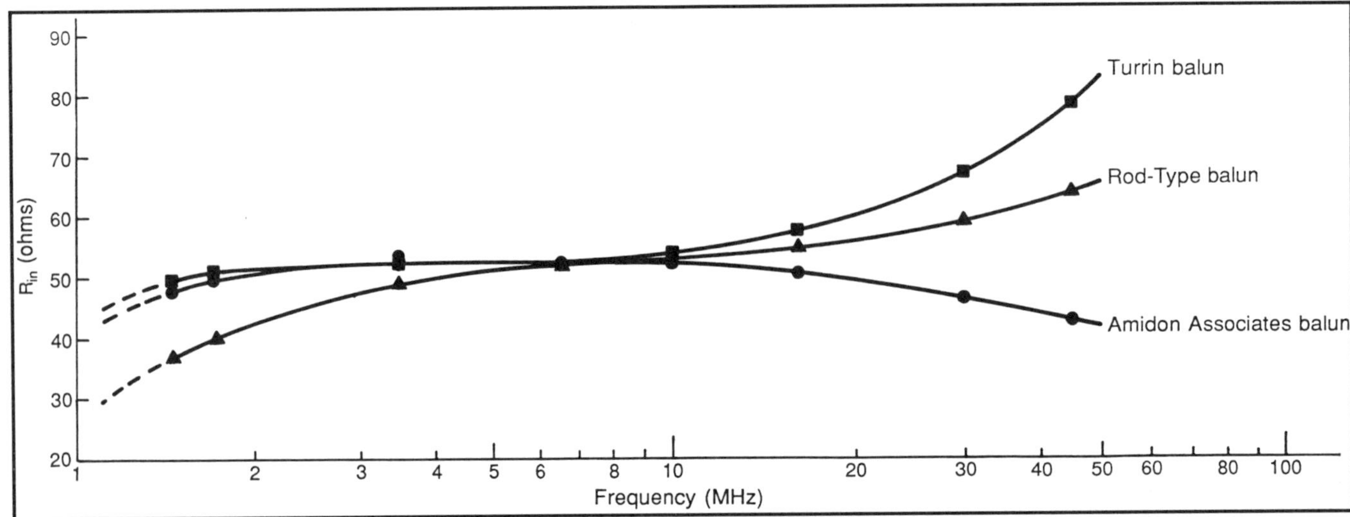

Figure 1-4. The input impedance versus frequency, when terminated with 50 ohms, for the Turrin, typical rode type, and Amidon Associates 1:1 baluns.

showed a drop in the input impedance and an inductive component at 2 MHz. This meant flux in the cores and an undesirable condition—especially for ferrite, which is a nonlinear material. Ferrite cores could not only suffer damage, but they could also generate spurious frequencies under these conditions. In fact, the same condition could occur at 4 MHz with a VSWR of 2:1! Therefore, I don't recommend any of these baluns for use on 160 or 80 meters.

2. The major problem at the low-frequency end is the role of the third winding (3-4) in **Figure 1-3B**. It has been claimed[12] that the third winding improves the low-frequency response (over the two-conductor Guanella 1:1 balun) because it enables autotransformer action at the low end. However, recent measurements I have made on two-conductor Guanella baluns and three-conductor Ruthroff (or Turrin) baluns, with loads grounded at their centers, show insignificant differences. This type of load approximates the actual condition when feeding a balanced antenna system. The negative feature of the third winding (3-4) is that, at the low-frequency end, there can be insufficient reactance to prevent harmful flux in the core because of a direct shunting path to ground. With the two-conductor 1:1 balun, the only flux-inducing current is that of the imbalance current (the inverted L mode), which is usually far smaller.

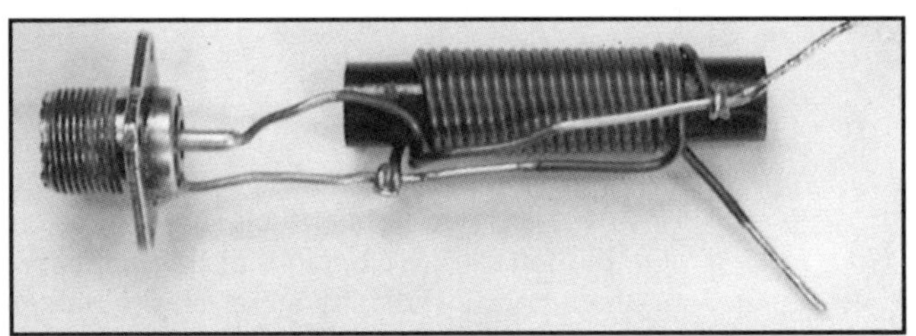

Photo 1-C. A typical rod-type balun.

3. Another important feature of the curves shown in **Figure 1-4** is the effect of the characteristic impedances of the coiled transmission lines. For example, a bifilar winding (wires tight together) on a toroid with spacing between adjacent bifilar turns exhibits a characteristic impedance of about 45 ohms. When wound on a rod with no space between adjacent bifilar turns, the characteristic impedance drops to about 25 ohms. With the third winding (3-4) between the other two as shown in **Figure 1-3B**, the characteristic impedance is raised to approximately 70 ohms in the toroidal case and to about 47 ohms in the rod case. If the toroidal baluns were terminated in 70 ohms and the rod balun terminated in 47 ohms, the high-frequency responses would be practically flat to at least 30 MHz. The difference in high-frequency response between the two toroidal baluns (with 50-ohm loads) is due to the differences in the lengths of their transmission lines. The transmission line on the powdered-iron core is appreciably less because the OD, as well as the cross-sectional area, is smaller.

4. The trifilar-wound form of the 1:1 balun also has an additional undesirable property. Its high frequency response is sensitive to unbalanced and mismatched loads. This is because the third wire now forms two tightly coupled transmission lines. It is unlike the Ruthroff version shown on the right in **Photo 1-A**. In

his second article,[12] Turrin pointed out this important distinction.

J. Reisert, W1JR—1978

The next significant article on 1:1 baluns was published by Reisert in 1978.[7] Reisert proposed winding some of the smaller (but still high-powered) coaxial cables around a 2.4-inch OD ferrite toroid with a permeability of 125. The windings also included a crossover, which is shown in **Figure 1-5** and **Photo 1-D**. In addition, he recommended various numbers of turns, depending upon the low-frequency requirement. For example, he suggested 12 turns to cover 3.5 MHz, 10 turns for 7 MHz, 6 for 14 MHz, and 4 for 21 and 28 MHz. Because the characteristic impedance of the coaxial cable is the same as the coax feedline, the balun only introduces a foot or two of extra length to the feedline. This is true in the HF and VHF bands. The coaxial cables recommended in the article were RG-141/U, RG-142/U, and RG-303/U.

From the articles that followed in the amateur radio literature, it became apparent that few recognized all of the important features of Reisert's balun. They are listed below.

1. An efficient, low-loss ferrite was used.
2. The baluns had sufficient choking reactances for the various low-frequency requirements.
3. The characteristic impedance of the coiled transmission line was the same as that of the feedline, eliminating the extra transformer action of a length of transmission line with a different characteristic impedance.
4. The balun is a form of Guanella's two-conductor 1:1 balun, which is not prone to core flux and, hence, saturation and the generation of spurious frequencies. It is also not susceptible to mismatched and unbalanced loads as are the Turrin and "voltage" baluns.

After constructing several of Reisert's baluns and comparing them with other Guanella designs, I found that the crossover winding had virtually no effect up to 100 MHz (the limit of my equipment). In regards to Reisert's VSWR comparison with a rod-type balun when feeding a triband Yagi beam on 20 meters, I found that his balun had a lower VSWR (practically 1:1) at the best match point. The rod-type balun had a best VSWR of about 1.3:1, but at a slightly higher frequency. He attributed the higher (and somewhat flatter) VSWR curve of the rod-type balun to its greater ohmic loss. Because the rod-type baluns I have investigated used the same low-loss ferrite that Reisert's did, I suspect that the differences in the VSWR curves were mainly due to the mismatch loss introduced by the rod-type balun.

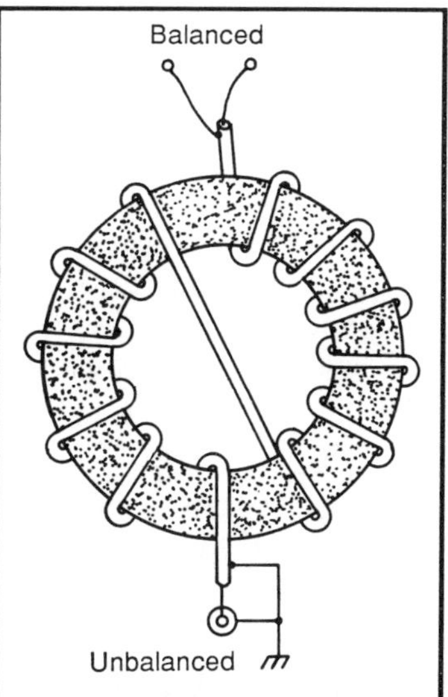

Figure 1-5. A pictorial of the crossover used in Reisert's 1:1 balun.

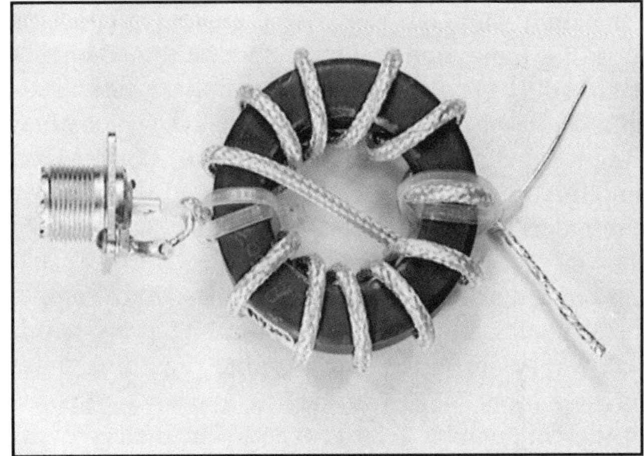

Photo 1-D. A Reisert, W1JR, 1:1 balun.

G. Badger, W6TC—1980

Badger published an in-depth, two-part series[13,14] in *Ham Radio* magazine in 1980 on air-core baluns and ununs. I am sure it was instrumental in advancing the technology of this class of wideband transformers. A recent article by Orr[15] also shows that there are many other radio amateurs who see the advantages of air-core transformers.

What are the claims for air-core baluns over their ferrite-core counterparts? First and foremost, proponents of air-core baluns claim they don't suffer the consequences of saturation that lead to spurious frequencies, heating, and ultimate damage. Secondly, it's said that they are not subject to arcing from the windings to the core.

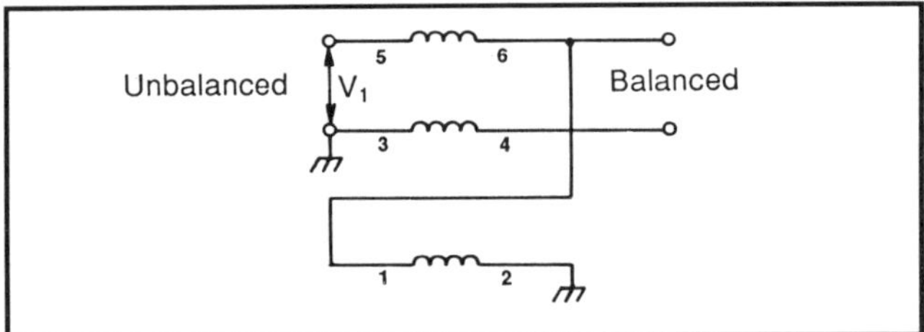

Figure 1-6. Schematic of Badger's 1:1 balun with a compensating winding (1-2). Winding (3-4) is the outer-braid of the coax and winding (5-6) is the inner conductor.

What are the claims for the ferrite-core baluns over their air-core counterparts? Simply put, they have wider bandwidths and are more compact.

After reading Badger's two-part series, I found that my curiosity was piqued by his experimental data on harmonic distortion due to saturation in a ferrite-core 1:1 balun. Although many have expressed concerns regarding saturation in ferrite-core baluns, Badger's data could very well provide the only results available. He used the two-tone test method, which combined two RF sources of 2.001 and 2.003 MHz, amplified it to 2 kW PEP and then fed it through a commercial 1:1 rod-type ferrite balun. The data showed considerable distortion in the 3rd order and 9th order distortion products. In other words, appreciable nonlinearity took place at this high power level.

Several questions come to mind regarding these measurements. What was the low-frequency response of the commercial 1:1 rod-type balun Badger used? From my measurements on a rod-type balun (**Figure 1-4**), I found a drop in the input impedance and an inductive component at 2 MHz. This indicates flux in the core and a problem when using this balun at 2 MHz. Because many rod-type 1:1 baluns have been used over the years, it would have been instructive if he had also made these measurements at 4 and 7 MHz. They would have given the readers a safe lower-frequency limit for these baluns.

I wondered why Badger didn't make similar measurements on Reisert's 1:1 balun, which he included in his articles. As noted earlier, I consider Reisert's 1:1 balun a very good design! I am sure that no distortion products would have been found at 2 MHz with this balun. The end result is that Badger chose a poor ferrite-core design for making his comparisons. This helped to contribute to an undeserved reputation for the ferrite-core balun.

Badger also suggested placing an insulated wire in parallel with the coax winding on Reisert's 1:1 balun. He called this a compensating winding, which provided a superior balanced output. The third winding (1-2) is shown in **Figure 1-6**. Winding (5-6) is the inner conductor and winding (3-4), the outer braid. Later experiments by myself and others have shown that a well-designed two conductor (Guanella) 1:1 balun has a completely satisfactory balanced output for antenna applications. Furthermore, it does not suffer from an unbalanced and/or mismatched load and core saturation. Incidentally, Badger's schematic of **Figure 1-6** now adds up to four different versions of the 1:1 balun. They are the two-conductor version of Guanella, and the three, three-conductor versions of Ruthroff, Turrin, and (now) Badger.

Badger and Orr also mentioned the Collins balun in their articles. This balun is comprised of a dummy length of coax wound as a continuation of the original coiled coax winding. Interestingly, it is connected as a Ruthroff 1:1 balun (**Figure 1-2B**), which also uses a third winding. Because there is appreciable coupling between the two coiled windings, the Collins balun should also be susceptible to mismatched and/or unbalanced loads. Badger claimed it was, by far, the best 1:1 balun he had ever used. Again, it would have been very informative if he had compared it with the Reisert balun (without the compensating third wire).

M.W. Maxwell, W2DU—1983

One of the more significant articles on 1:1 baluns was published by Maxwell[4] in 1983. Maxwell introduced, what he called, the "choke" balun. It was formed by placing high-permeability ferrite beads

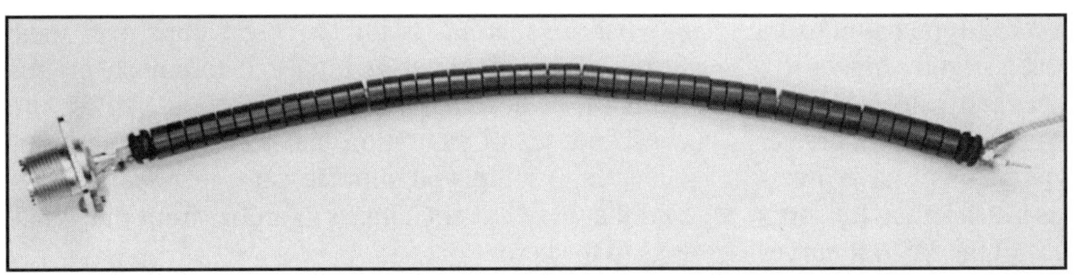

Photo 1-E. The Maxwell, W2DU, "choke" 1:1 balun.

over about one foot of small (but high-powered) coaxial cables similar to the ones used in the Reisert balun. **Photo 1-E** shows the W2DU "choke" balun removed from its plastic enclosure.

Maxwell compared his balun with (what he termed) a "transformer-type" balun by measuring the input impedances versus frequency when the outputs were terminated in 50 ohms. The "transformer-type" balun didn't yield a true 1:1 impedance transfer ratio, which he claimed was due to losses, leakage reactance, and less than optimum coupling. Because Maxwell gave no description of the "transformer-type" balun, I assumed it was the popular rod-type balun shown in **Photo 1-C**. As you can see in **Figure 1-4**, this balun has a poor low-frequency response. Furthermore, it is really optimized for a load of 47 ohms, not 50 ohms.

What Maxwell failed to realize was that his balun was a form of Guanella's two-conductor type. That is, it is both a choke (a lumped element) and a transmission line (a distributed element). Additionally, Guanella's theory applies whether the transmission lines are coiled (about a core) or beaded, twin-lead or coaxial cable. From Ruthroff's classic paper,[9] which extended Guanella's work,[3] we became aware of the voltage drops along the lengths of the transmission lines. From very accurate insertion loss measurements,[2] we learned that the losses were mainly in the magnetic medium—and that they were related to the voltage levels and the permeabilities. Maxwell didn't take into account these latter findings. He used lossy high-permeability beads (2500) and assumed that the main loss was in the transmission line. He claimed that the CW power-handling capability of his balun was 3.5 kW at 50 MHz and 9 kW at 10 MHz—the same as the coaxial cable itself. I seriously question these power ratings. Ironically, it is very likely that Maxwell's balun had more real loss than the so-called "transformer-type" balun!

R.W. Lewallen, W7EL—1985

There is very little doubt that Lewallen's interesting article[6] in 1985 contributed significantly to the better understanding and design of 1:1 and 4:1 baluns. In it, he coined the (now very popular) terms "voltage" and "current" baluns. The "voltage" balun, a three-conductor type, has output ports which have voltages that are balanced to ground. It is brought about (see **Figure 1-6**) by the voltage-divider action of windings (5-6) and (1-2). Because we have two tightly coupled transmission lines in the passband with the same potential gradients, terminal 6 is at $+V_1/2$ and terminal 4 at $-V_1/2$ where V_1 is the input voltage. The "current" balun, on the other hand, is a two-conductor balun that produces equal and opposite currents on the output ports for any form of load impedance.

Lewallen conducted a series of experiments on 10 meters to compare the performances of "voltage" and "current" baluns under balanced and unbalanced conditions. In the unbalanced (nonsymetrical) condition, the dipole was lengthened by five inches on one side and shortened by five inches on the other side. He then obtained a figure of merit for both baluns (as well as for the case without a balun) defined as the ratio of the average magnitude of the currents at the feedpoint over the magnitude of the imbalance (the inverted L) current. The magnitudes of the currents were obtained by current-probe toroids. Measurements were made at the antenna feedpoint and at a half-wave (physically) from it.

The "current" balun consisted of 15 turns of very small RG-178/U coax on a FT82-61 core (a ferrite toroid with an OD of 0.825 inches and a permeability of 125). The "voltage" balun had 10 turns of RG-178/U coax with a No. 26 wire in parallel (closely coupled) on the same toroid. The schematic is shown in **Figure 1-6**.

Lewallen concluded (and I agree) that his experiments clearly showed that the "current" balun gave superior performance at every measured point in each experiment. However, the "voltage" balun still improved the balance over the no-balun case. He also concluded that other experiments should be performed in order to better compare the two forms of the balun. One is to ascertain the difference when the feedline is placed nonsymetrical with respect to the antenna (to induce an imbalance current into the feedline). Others include determining the optimum point in the feedline to place the balun, and testing the various kinds of core and beaded baluns.

Although Lewallen's article pretty much speaks to Badger's proposal of adding a third wire to Reisert's balun for better balance (i.e., avoid it), there are some comments and questions I have regarding his experiments and findings. They are:

1. Why didn't Lewallen use Reisert's balun as the "current" balun and Badger's suggested third-wire design as the "voltage" balun? These would have been more realistic designs for comparisons. Instead, he used very small structures (which will only find use in QRP operations) and, as such, have higher frequency capabilities. Also, because the "voltage" balun only had 10 turns (and hence a shorter transmission

line and a poorer low-frequency response), it was favored in the comparisons on 10 meters. Had Lewallen used transmission lines of equal lengths on Reisert's cores, the differences between the two baluns would have been even more dramatic.

2. It would also have been very useful if Lewallen had made comparisons between a "current" balun that could handle the full legal limit of amateur radio power (again like the Reisert balun) and "voltage" baluns using Turrin's schematic (**Figure 1-3**) with rod and toroidal cores which have been readily available for nearly three decades.

3. Additionally, comparisons should not only be limited to 10 meters. Because 1:1 "voltage" baluns are configurations of coupled transmission lines with various characteristic impedances, their performances with mismatched and unbalanced loads are more sensitive to the higher frequencies than their "current" balun counterparts. Therefore, making similar measurements on 20 meters would also provide more useful information.

4. Even though Ruthroff's classic 1959 paper[9] has been the industry standard over the years, his 1:1 balun design has been practically nonexistent in the amateur literature. Turrin mentioned its advantage over his first design in his second article.[12] But Turrin's first design has prevailed in our amateur literature. Because Ruthroff's design has the third conductor on a separate part of the toroid, it has the balanced output mentioned by Badger,[13] but still retains the flexibility of the Guanella[3] balun. In other words, as the frequency is increased, the choking action of the third wire makes it practically transparent. This enables it to handle any form of load impedance. It would have been informative if Lewallen had pointed this out and also noted that Ruthroff's 1:1 balun, although looking like a "voltage" balun, is really a "current" balun.

5. Lewallen and the others who have published in the amateur radio literature have failed to reference the first presentation on what are now known as "current" or "choke" baluns, made by Guanella[3] in 1944. Even though Guanella used coiled transmission lines without a magnetic core, his theory on how these devices work is still applicable today.

J.S. Belrose, VE2CV—1991

The last article on 1:1 baluns I consider worth mentioning was written by Belrose[5] in 1991. In it, he described the W2DU balun by Maxwell and how his technique of threading coaxial cable through ferrite beads could be easily applied to 4:1 and 9:1 baluns.

What immediately grabbed my attention in this article was the *deck head* (the editor's comments), which included highly complimentary remarks regarding the beaded-coax balun. In essence, it said, "In this breakthrough article, W2DU's peerless 1:1 current-balun design serves as the basis for excellent ferrite-bead-choke current baluns capable of 4:1 and 9:1 impedance transformation."

However, if one reads the article carefully, it becomes apparent that this is not what Belrose said. His words were, "The *current balun* of the *type* developed by Walt Maxwell, W2DU—a balun consisting of ferrite beads slipped over a length of coaxial cable—is the best so far devised (italics mine)." He did not say that W2DU's balun was "peerless." In fact, in the article he said just the opposite. He pointed out that the W2DU balun's main disadvantage is that the beads are lossy at HF and heating becomes a concern when the transmitting power exceeds 125 watts! For high power (that is, 1 kW CW), Belrose recommended Roehm's[16] designs, which use lower permeability (850) beads nearest the balun's balanced output (where most of the heating takes place).

However, I do question two of the advantages he claims for the W2DU balun. They are:

1. Its excellent power-loss and impedance-versus-frequency characteristics are much superior to those of a bifilar current balun wound on a ferrite toroid.

2. It has excellent power-handling capability, and can function quite satisfactorily when working into highly reactive loads. This is so because the magnetic flux produced by currents flowing on this balun's wires cannot saturate its ferrite beads.

Belrose obtained evidence for advantage number 1 by comparing the input impedance and power loss versus frequency of the W2DU balun with a commercial balun when they were terminated in 50 ohms. The commercial balun was a bifilar wound toroidal type used in a differential-T tuner. What Belrose failed to realize was that the commercial balun had heavily insulated wires, resulting in a characteristic impedance greater than 100 ohms. Thus, he was actually comparing a 50-ohm transmission line with a longer line that had a characteristic impedance in excess of 100 ohms! As expected, his input impedance versus frequency curve for the commercial balun was even more severe than that of the Turrin balun shown in **Figure 1-4**.

Advantage number 2 is based upon the premise that the magnetic flux produced by currents on the W2DU

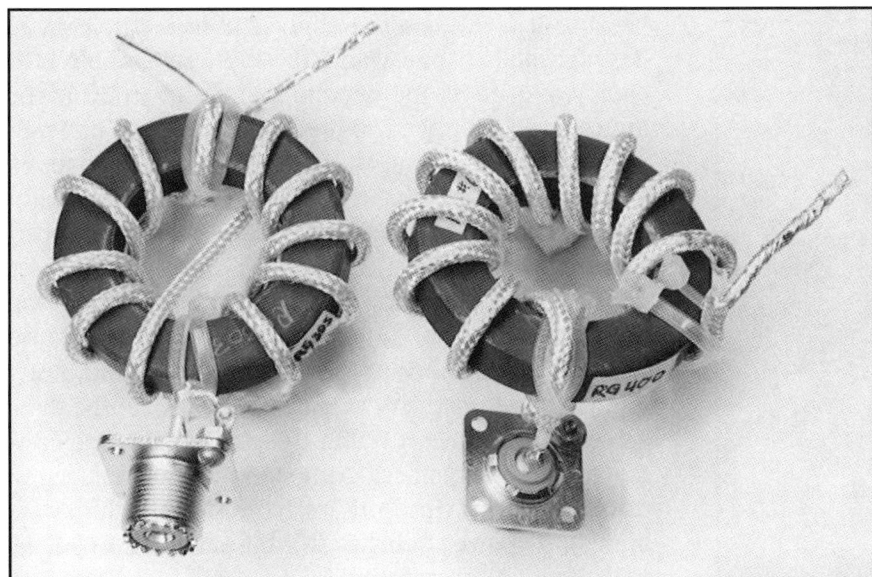

Photo 1-F. Two versions of Reisert's 1:1 balun. The balun on the left uses the crossover shown in Figure 1-5. The balun on the right is continuously wound. Both have the same electrical performance in the HF band.

balun's wires cannot saturate the ferrite beads, while the windings of a bifilar wound toroidal current balun can. This is an incorrect assumption because the magnetic flux of a two-conductor type balun, like the beaded-coax or the bifilar-wound toroidal balun, is generated by the imbalance (inverted L) current and hence is *much* lower than the transmission line currents. This is especially true with sufficient choking reactances. The perception that the toroidal type balun still transmits the energy to the output circuit by flux linkages could well lead to this mistaken impression.

For high power beaded-coax baluns, Belrose referred to designs by Roehm,[16] who increased the power capability of this type balun by using lower permeability beads near the balanced output. He also increased the length considerably. For operation from 80 meters to 10 meters, Roehm used 28 inches of beaded coax. For 160 meters to 10 meters, he used 36 inches of beaded coax. Belrose's suggestion of connecting beaded coaxes in parallel on the low-impedance side and in series on the high impedance side to obtain a broadband 4:1 transformation ratio would require transmission lines with characteristic impedances of 100 ohms. This means, for a high power 4:1 balun using beaded transmission lines, about 56 inches of beaded line would be required for the 80 meter to 10 meter operation and 72 inches for the 160 meter to 10 meter coverage. For a 9:1 balun, it would be necessary to increase these lengths by 50 percent!

It remains to be seen what Belrose would have said or done if he had compared the W2DU balun of Maxwell's with the W1JR balun of Reisert's. He certainly couldn't claim the advantages listed in his article for the W2DU balun. Would he still have claimed that the type of balun developed by Maxwell is the best so far devised? Given the evidence, I doubt it.

Sec 1.4 High-, Medium-, and Low-power Designs

In this section I'll present my latest 1:1 balun designs. Except for one balun that appeared in the June 1993 issue of *CQ*, the others appeared for the first time in the magazine's April 1994 issue. Because I have favored Reisert's design throughout this chapter, the first baluns described here are my versions of his technique of coiling small (but high power) coaxial cable around a low-permeability ferrite toroid. For my wire versions, I could have used all sorts of adjectives to describe them like Guanella, two-conductor, choke, and current. However, in the process of writing this section, I thought Belrose's adjectives were the most direct. Using his words, I call my wire versions of the 1:1 balun simply—*bifilar toroidal baluns.*

Photo 1-F shows two versions of Reisert's balun. The one on the left uses the crossover shown in **Figure 1-5**. Because no difference in performance at HF was noticed without the crossover, a continuous-wound version is also shown on the right. The main advantage in the HF band with the crossover winding is purely mechanical. Having the input and output connections on opposite sides of the toroid is not only more convenient, but it also offers a much stronger method of mounting.

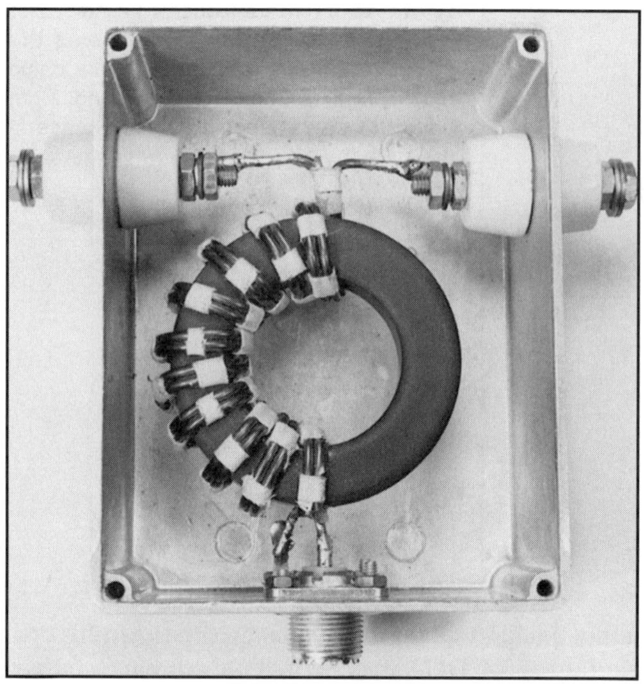

Photo 1-G. My high-power design of a bifilar toroidal (Guanella/current) 1:1 balun mounted in a 4 inch long by 3 inch wide by 2.25 inch high Bud aluminum box.

For operation from 1.8 to 30 MHz, 10 turns of small coax like RG-303/U, RG-142B/U, or RG-400/U are wound on a 2.4 inch OD ferrite toroid with a permeability of 250. If the use is limited from 3.5 to 30 MHz, then a permeability of 125 is recommended because it would yield a slightly higher efficiency at the high end. If one wants the highest possible efficiency and limits the operation from 14 to 30 MHz, then a permeability of 40 is recommended. With loads grounded at their centers, these conditions were found to give ample margins (handle a VSWR of 3:1 without any appreciable flux) at their low frequency ends.

For ease of winding, I found TY-RAP™ Cable Ties *very* useful. Two were used at each end. Removing the covering on the outer braid also helps. Because about 24 inches of cable is wound on the toroid, I recommend you start with at least 32 inches. Of the three cables noted above, I found RG-303/U cable the easiest to wind and connect. Although it only has a single-thickness braid (the others have double-thickness braids), its power rating is still the same—9 kW at 10 MHz and 3.5 kW at 50 MHz.

The next high power design is shown in **Photo 1-G** mounted in a 4 inch long by 3 inch wide by 2.25 inch high Bud CU 234 aluminum box. It has 10 bifilar turns of No. 12 H Thermaleze wire on a 2.4 inch OD ferrite toroid. As with the Reisert versions, a permeability of 250 is recommended for 1.8 to 30 MHz, 125 for 3.5 to 30 MHz, and 40 for 14 to 30 MHz. One wire is also covered with two layers of Scotch No. 92 polyimide tape in order to raise the characteristic impedance to 50 ohms. With this added insulation, the voltage breakdown of this twin-lead transmission line compares very favorably with RG-8/U cable (4000 volts). In order to preserve the spacing, the wires are

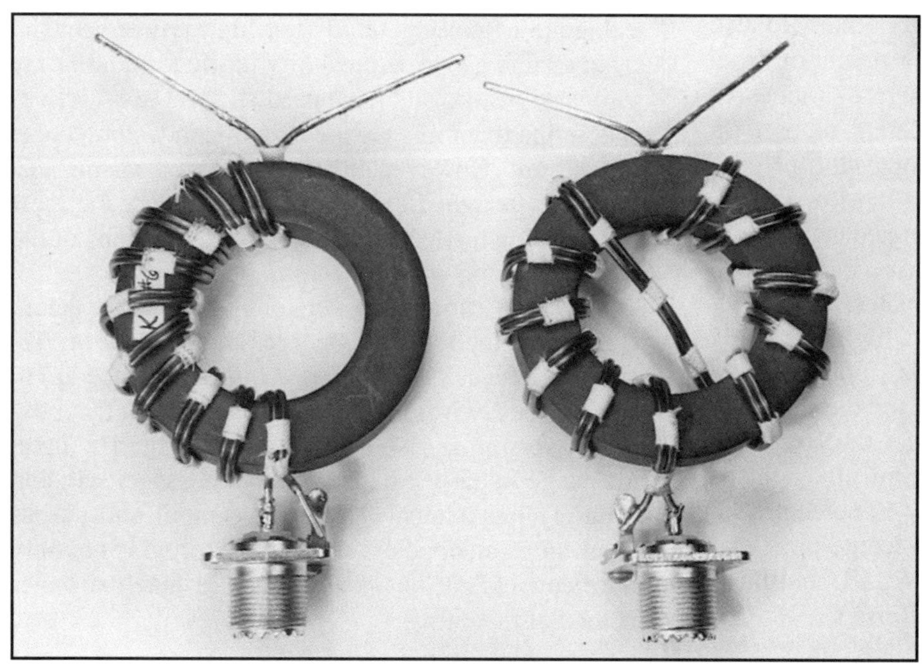

Photo 1-H. Two "economy" versions of the high-power bifilar toroidal (Guanella/current) 1:1 balun. The one on the right uses Reisert's crossover technique.

also clamped together about every 1/2 inch with strips of Scotch No. 27 glass tape 3/16 inches wide and a little over 1 inch long.

Two "economy" versions of the high power bifilar toroidal balun are shown in **Photo 1-H**. The one on the left shows the windings crowded on one-half of the toroid. The one on the right provides the same positions of the input and output connections by using the crossover. Their performances are identical. Both baluns have 10 bifilar turns of No. 14 H Thermaleze wire on a 2.4 inch OD ferrite toroid. The choices of permeability, which trade off bandwidth for efficiency, are the same as those used in the two previous high power designs. "Economy" refers to economy in labor. This balun is actually very easy to construct.

This "economy" balun, which also handles the full legal limit of amateur radio power, has a small tradeoff in high frequency response. Because no extra insulation is used, the characteristic impedance of two tightly clamped No. 14 H Thermaleze wires is 45 ohms. With one layer of Scotch No. 92 tape, it increases to 50 ohms. But for most of the HF band, the difference in performance between baluns using transmission lines of 45 and 50 ohms should be negligible. Even without the extra insulation, the voltage breakdown should compare very favorably with the smaller, high-power coaxes used in the Reisert versions (1900 volts).

Photo 1-I shows two low-power versions of a bifilar toroidal balun capable of handling the output of practically any HF transceiver. One has a crossover winding and the other, a continuous winding. They both have 10 bifilar turns of No. 16 H Thermaleze wire on a 1.25 inch OD ferrite toroid with a permeability of 250. Since efficiency is not a major problem in low-power use, I found no reason to suggest the other two versions which use lower permeabilities. It is also interesting to note that two tightly clamped No. 16 H Thermaleze wires have a characteristic impedance close to 50 ohms. Therefore, this small balun (particularly with its short leads) has a very good high frequency response.

Photo 1-J shows two medium-power versions of a bifilar toroidal balun capable of handling the full legal limit of amateur radio power under controlled conditions—when the VSWR is less than 2:1. Being smaller than its larger (2.4 inch OD) counterpart, its heatsinking capability and hence, power rating is less. As before, one balun uses a crossover while the other doesn't. Each has 8 bifilar turns of No. 14 H Thermaleze wire on a 1.5 inch OD ferrite toroid. The ferrite permeabilities and expected bandwidths are the same as the other high-power baluns. Because the average magnetic path length in the core is about two-thirds that of the 2.4 inch core, only 8 bifilar turns are required in order to produce a similar low-frequency capability. Even though the characteristic impedances of their bifilar windings are 45 ohms, their responses on 10 meters should be somewhat better than the "economy" models because the lengths of their trans-

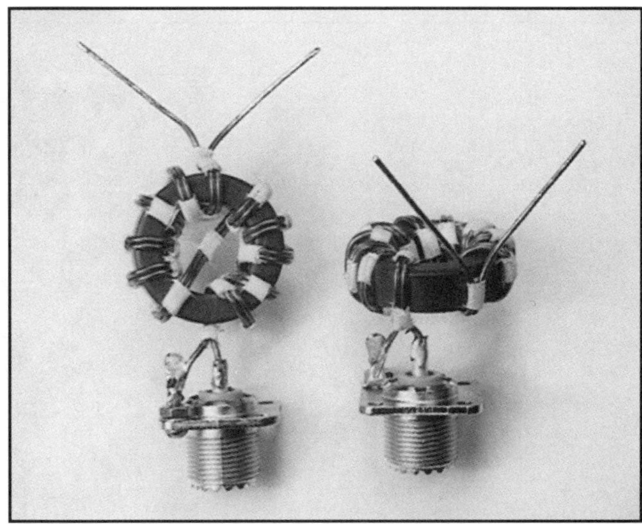

Photo 1-I. Two low-power versions of the bifilar toroidal (Guanella/current) 1:1 balun capable of handling the output of practically any HF transceiver. The balun on the left has the crossover.

Photo 1-J. Two medium-power versions of the bifilar toroidal (Guanella/current) 1:1 balun capable of handling the full legal limit of amateur radio power when the VSWR is less than 2:1. The balun on the left has the crossover.

Photo 1-K. Two examples of mounting the high-power 1:1 Guanella balun in enclosures that allow isolation transformer operation.

mission lines are shorter (18 compared to 24 inches).

And now a few words on what sort of efficiency one can expect in trading-off low frequency response by using lower permeability ferrite cores. From earlier studies,[2] it was found that the efficiency (with sufficient choking, so only transmission line currents flow) is related to the permeability, the voltage drop along the length of the transmission line, and the frequency. The higher the permeability and/or voltage drop, the greater the loss. Additionally, the higher the permeability, the greater the loss with frequency. It was also found that a permeability of less than 300 was necessary in order to obtain the very high efficiencies of which these devices are capable.

Here are some efficiencies that might be expected from ferrites under matched conditions, based on the results of the studies:

1. With 250 material, an efficiency near 99 percent at 1.8 MHz and 97 percent at 30 MHz.
2. With 125 material, an efficiency near 99 percent at 3.5 MHz and 98 percent at 30 MHz.
3. With 40 material, an efficiency of 99 percent at 14 MHz and at 30 MHz.

When a balun is exposed to a high impedance resulting in a VSWR of 2:1, the voltage, and hence loss, increases by about 40 percent. With a VSWR of 4:1, the loss doubles. With a VSWR of 10:1, the loss is more than three-fold. Since limited data was obtained in this study,[2] these increases in losses with increases in VSWR could very well be even greater.

Sec 1.5 Isolation Transformers

As was shown in the preceding sections, the 1:1 Guanella (current/choke) balun is in reality a two-conductor RF choke, or an isolation transformer. However, practical balun designs don't usually lend themselves to isolation use for mechanical reasons. They cannot be conveniently inserted into a coaxial cable system. Therefore, the objective is to mount the "Guanella Baluns" in enclosures that will allow them to perform as isolation transformers.

Photo 1-K shows two examples of enclosures that should accomplish the mission. The example on the left shows one of the high-power units mounted in a 5 inch long by 3.5 inch wide by 2.25 inch high aluminum enclosure. The output connector (on the top) is insulated by a 0.125-inch thick piece of plastic. The one on the right shows another high-power unit mounted in a 6 inch long by 3 inch wide by 2 inch high plastic enclosure, which is available from Radio Shack. The descriptions of the high-power units are given in the preceding section.

These isolation transformers can be used as baluns when inserted in a coaxial cable one-half wavelength (physically) from a half-wave dipole[6] or between a coaxial cable and a balanced L-C tuner.[16] They should never be inserted in a coaxial cable system that presents a high impedance and, consequently, a high voltage. This could be harmful to the unit.

Note that the high-power unit on the left in **Photo 1-K** can also be used as a filament choke in a Class B linear amplifier. Its low-frequency response is much better than any rod-type, assuring 160-meter operation. Obviously, it would not be mounted in the aluminum enclosure shown in the photo. In addition, the two layers of Scotch No. 92 tape would not be required because the characteristic impedance of the winding is unimportant and the voltages involved are quite low.

By the way, the designs discussed above can be used as phase-inverters. Simply connect the hot lead on one side to the ground on the other side.

Sec 1.6 Summary

In investigating the 1:1 balun, I was quite surprised to see the ferrite- and powdered-iron-core designs that have been available in the literature and off-the-shelf since 1964. They not only had poor low- and high-frequency responses, but they were also susceptible to flux in the cores at their low-frequency ends. Furthermore, since they only used single-coated wires, they were also prone to voltage breakdown. There is no doubt that these designs are responsible for the poor reputation that the balun has had for many years.

It wasn't until 1978, when Reisert published his article, that a balun which had all of the attributes of a good design became available. Namely:

a) It is efficient because it uses a low-permeability core.

b) It has sufficient choking reactance to meet its low-frequency requirement.

c) It is not prone to flux in the core (and hence, saturation) because it has no third winding.

d) It has a 50-ohm characteristic impedance and maintains a 1:1 transformation ratio with a 50-ohm load.

e) It has a good voltage breakdown capability (1900 volts).

f) It can handle a mismatched and/or unbalanced load.

However, succeeding investigators failed to see the advantages of Reisert's design and proposed their own. Surprisingly, they belonged to two distinct groups. One group favored "air-core" baluns and the other, "choke" (beaded-coax) baluns.

The main argument given by the "air-core" followers was that their balun would never experience problems with saturation, while the "ferrite-core" balun would. However, the Reisert balun is a current/choke type balun that could only have flux in the core due to the imbalance (inverted L) current, which is much smaller than the transmission line currents. In fact, with any degree of choking reactance by the coiled transmission line, the imbalance current is essentially negligible. Therefore, saturation is not a concern with a balun like Reisert's. In all fairness, it should be noted that it is a different story with the 4:1 current/choke and voltage baluns. All three of these types of baluns have a "magnetizing inductance" in their low-frequency models and hence a possibility of saturation with a poor design.

The advocates of the "choke" 1:1 balun claim that their beaded-coax balun can't saturate while the bifilar (current) toroidal balun can. This is untrue; they are basically the same kind of structure, and neither has a third conductor which could allow a flux-causing current at the very low-frequency end. But of all of the attributes listed above for the Reisert balun, the first one has the "choke" balun at a disadvantage in the HF band. Because its transmission line is not coiled about a toroid, it does not have the multiplication factor of N^2 (due to mutual coupling), where N is the number of turns, while the toroidal balun does. Therefore, higher-permeability beads are required in order to obtain sufficient choking reactance. This results in lower efficiency.

Finally, it should be pointed out that the 1:1 balun using the small, but high-power coaxial cable, is capable of 5 kW operation in the HF band. Because of current crowding, the bifilar toroidal balun even with No. 12 wire has shown excessive wire heating at this level. Therefore, No. 10 or No. 8 wire, with added layers of Scotch No. 92 tape in order to obtain a 50-ohm characteristic impedance, is recommended.

Chapter 2

The 4:1 Balun

Sec 2.1 Introduction

The first chapter of this book discussed the most popular of all baluns—the 1:1 balun designed to match 50 ohms unbalanced to 50 ohms balanced. It not only gave a review of the history, theory, and design of these broadband transformers, but also my viewpoint on recently published articles advocating "new" designs using coaxial cable wound around a toroid, threaded through ferrite beads, or just plain coiled in air (an air-core balun). As was noted, I am in considerable disagreement with many of the claims advanced for these "new" 1:1 baluns.

Chapter 2 deals with the next most popular balun—the 4:1 balun designed to match 50 ohms unbalanced to 200 ohms balanced. Baluns matching 50 ohms unbalanced to 12.5 ohms balanced are also included. It begins with a little information on the history and design of these baluns, followed by high- and low-power designs and comparisons with other baluns that have been on the market or in the amateur literature. Guanella's approach to the design of a 4:1 balun is particularly noteworthy. It not only yields an excellent balun design, but also an unun (unbalanced-to-unbalanced transformer) with practically the same performance. His design could very well be called a *balun/unun*. As in the first chapter, this one also closes with a brief summary of the significant points.

Sec 2.2 A Little History and Design Information

There are really only two classic papers that have established the principles upon which the transmission line transformer (the balun being a subset thereof) is based. The first one was written by Guanella in 1944. Guanella proposed the idea of coiling a transmission line to isolate the input from the output, resulting in the (now popular) *current* or *choke* balun.[3] The second was by Ruthroff in 1959, whose analysis of these transmission line transformers is the present industry standard.[9] Ruthroff also introduced the unun and the hybrid transformer.

Interestingly enough, both Guanella and Ruthroff had different approaches to their 1:1 and 4:1 balun designs. Guanella used a two-conductor 1:1 balun design, while Ruthroff used a three-conductor design. Ruthroff's third conductor (which was said to increase the low-frequency response over the two-conductor balun[12]) lay on a separate part of a toroidal core. Investigators who followed failed to recognize this fact. Their comparisons were made with a three-conductor balun that had the third wire in parallel with the other two, which then formed two coupled transmission lines. This gave rise to the term *voltage* balun—an inferior design.[6]

However, the differences between Guanella's and Ruthroff's approaches to 4:1 baluns were even more striking. Guanella connected coiled transmission lines in a parallel-series arrangement, so in-phase voltages were summed at the high impedance side. His balun has been called a *current* balun.[6] Ruthroff, on the other hand, obtained a 4:1 transformation ratio by summing a direct voltage with a delayed voltage that traversed a single transmission line in a phase-inverter connection (see **Chapter 1**). His balun has been called a *voltage* balun. The distinction between the operation of these two baluns was also overlooked by practically everyone who followed.

This section reviews the two different approaches taken by Guanella and Ruthroff in obtaining 4:1 baluns. Of particular importance are the descriptions of the potential drops along the lengths of the trans-

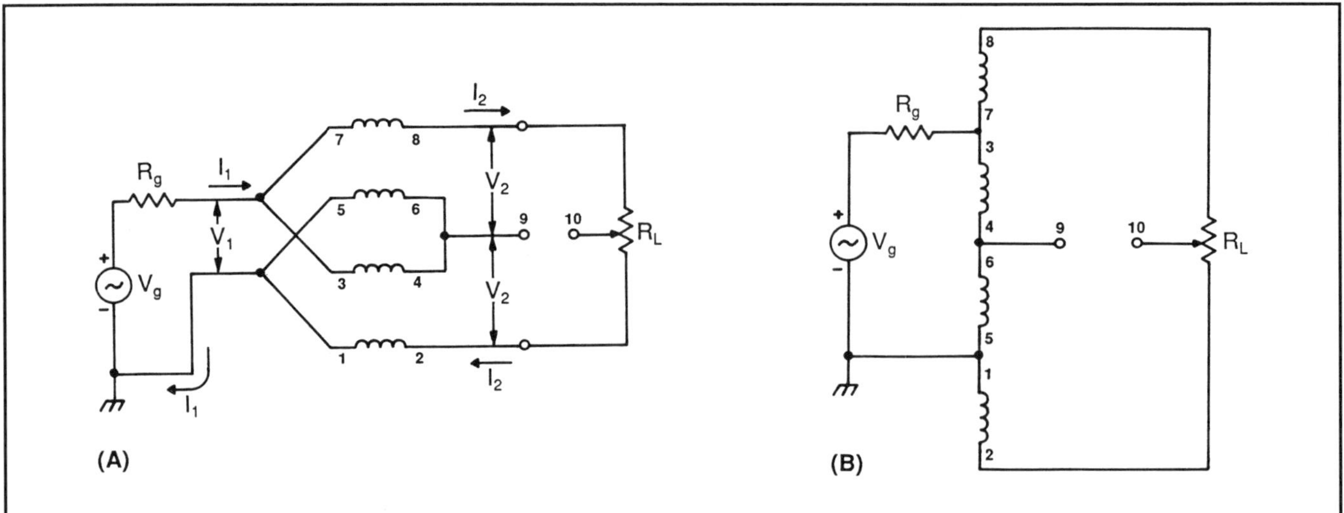

Figure 2-1. Electrical models of the Guanella 4:1 balun: (A) high-frequency, (B) low-frequency.

mission lines when the loads are grounded at various points. These voltage drops are not only relevant to the ohmic losses in the baluns, but also to their electrical performances. These descriptions were quite possibly presented for the first time in the second edition of my book *Transmission Line Transformers*.[2]

Sec 2.2.1 Guanella's 4:1 Balun

Figures 2-1A and **B** show the high- and low-frequency models of Guanella's method of connecting transmission lines in parallel-series to obtain a 4:1 balun. The high-frequency model (**Figure 2-1A**) assumes that the choking reactances of the coiled (or beaded) transmission lines are sufficient to isolate the input from the output, so only transmission line currents are allowed to flow. This occurs when the reactance of windings 3-4 and 5-6 (which are in series) is much greater than R_g (at least by a factor of ten[2]). If two cores are used, the reactance is the sum of the reactances of windings 3-4 and 5-6. If a single core is used, the reactance is twice as large because of the mutual coupling between the windings. The other advantage of Guanella's method (besides only using one core) is that shorter transmission lines can be used, resulting in better high-frequency performance.

As with all transmission line transformers, the objective is to have the transmission lines see loads equal to their characteristic impedances, resulting in "flat lines." This yields the highest frequency response. Because each transmission line in **Figure 2-1A** sees one half of the load, R_L, the optimum value of the characteristic impedance is $R_L/2$. Consequently, the input impedance, V_1/I_1, is simply the impedance of two identical transmission lines connected in parallel. It follows that the impedance transformation ratio is the load, R_L, divided by the input impedance.

Because the Guanella 4:1 balun sums voltages of equal delays from identical transmission lines, his balun is only limited in high-frequency performance by the deviation of the characteristic impedance of the transmission lines from the optimum values and the parasitics not absorbed into the characteristic impedance of the lines. I (and practically everyone else) had overlooked the simple and important statement, "a frequency independent transformation," which appeared in Guanella's 1944 article[3]—a fact that is evidenced by the scarcity of his designs in the literature. Another interesting aspect of the Guanella 4:1 balun is the analysis of his balun when the load is floating or grounded at different points. This leads to the determination of the voltage gradients that exist along the transmission lines and the various functions of which his 4:1 design is capable. Assuming a matched load or very short transmission lines resulting in $V_2 = V_1$, they are as follows.

Floating Load

With terminal 10 (which is at the center of R_L) floating, the potential gradient along the top transmission line in **Figure 2-1A** (windings 5-6 and 7-8) is $-1/2V_1$; along the bottom transmission line (winding 1-2 and 3-4) it's $-3/2V_1$. The voltage to ground on terminal 9, V_{90}, is $-1/2V_1$. Due to the fact that the bottom transmission line (in **Figure 2-1A**) has a voltage drop along its length three times greater than the top transmission line, it results in three times more loss because losses in transmission line transformers are voltage dependent (dielectric-type losses).[2]

Even though a single-core Guanella 4:1 balun maintains the voltages (stated above) when feeding a folded dipole (of about 200 ohms) which has a virtual-ground potential at terminal 10, it still feeds equal currents to each side of the antenna because of the series-connection at its output. Since the output voltages are not balanced to ground, a reactive component is probably introduced into the input impedance. Additionally, the choking reactance of the windings also prevents antenna currents from flowing on the outside of the coaxial cable feedline.

Load Grounded at Center

When two cores are used and terminal 10 (the center of R_L) is grounded, the voltage gradient along the top transmission line in **Figure 2-1A** is zero and along the bottom transmission line it is $-V_1$. The voltage to ground on terminal 9 (V_{90}) is also zero. In fact, the core for the top transmission line isn't needed. It merely acts as a mechanical support for the top transmission line, which now only operates as a delay line. Also, all of the loss now occurs in the core of the bottom transmission line where a longitudinal potential gradient exists. Furthermore, the low-frequency response, as seen from **Figure 2-1B**, is now determined by the reactances of windings 1-2 and 3-4. This means that the low-frequency response with a floating load is better by a factor of two over the case where the load is grounded at its center.

But the single-core case is a different matter. Since the potential at terminal 9 (V_{90}) wants to be at $-1/2V_1$, connecting a ground directly to the center of R_L causes an imbalance that renders the single-core balun unusable. If the ground were placed at a point 25 percent below terminal 8 (50 ohms from terminal 8 with a 200-ohm load), no difference would be noted from a floating load. This condition also exists when two cores are used.

Load Grounded at the Bottom

It is when the load is grounded at the bottom (at terminal 2), that we have what is probably the most interesting case. The 4:1 balun (with two cores) is now converted into a very broadband unun (unbalanced-to-unbalanced transformer). Because the bottom transmission line in **Figure 2-1A** has no potential drop along its length, it only acts as a delay line. The voltage to ground at terminal 9 (V_{90}) is $+V_1$, and the voltage gradient along the top transmission line is $+V_1$. This results in a voltage of $2V_1$ across the load. The low-frequency response is now determined by the reactances of windings 5-6 and 7-8. This is just the opposite of the balun case when the center of the load was grounded. A single-core 4:1 Guanella balun can also be converted to an unun by putting a 1:1 balun (for isolation) in series with the 4:1 balun.[2]

There is a reason for claiming a very broadband response for a Guanella unun (converted from his balun)—the two in-phase voltages are now summed at the high-impedance side. The only other competition for a 4:1 unun design is that of Ruthroff's,[9] where a direct voltage is summed with a delayed voltage that traversed a single transmission line (and, hence, had a built-in, high-frequency cut-off). In fact, very little information can be found in the literature on a *Guanella 4:1 unun*.

Sec 2.2.2 Ruthroff's 4:1 Balun

Figures 2-2A and **B** show the high- and low-frequency models of Ruthroff's approach for a 4:1 balun. The high-frequency model (**Figure 2-2A**) assumes that the choking reactance of the coiled (or beaded) transmission line is sufficient to isolate the input from the output, so only transmission line currents are allowed to flow. This occurs when the reactance of winding 3-4 (or 1-2, because they are the same) is much greater than R_g (at least by a factor of ten[2]).

As **Figure 2-2A** indicates, the transmission line is connected in a *phase-inverter* function (see **Chapter 1**). That is, a $-V_1$ voltage gradient now exists along the length of the transmission line. Therefore, the voltage across R_L now becomes $V_1 + V_2$. Although Ruthroff analyzed his 4:1 unun in his classic paper,[9] his results also apply to his balun with a floating load, because both devices sum a direct voltage with a delayed voltage. In essence, he used loop equations on the input and output and transmission line equations to eliminate one set of variables (I_2 and V_2). Ruthroff also used a maxima technique (setting a derivative to zero) to solve for the optimum characteristic impedance of the transmission line. As in the Guanella case, he found the optimum value to be $1/2R_L$.

An inspection of **Figure 2-2A** shows that the left side of R_L (terminal 3) has a direct voltage (V_1) to ground, and the right side (terminal 2) a delayed voltage ($-V_2$) to ground, which traveled the length of the

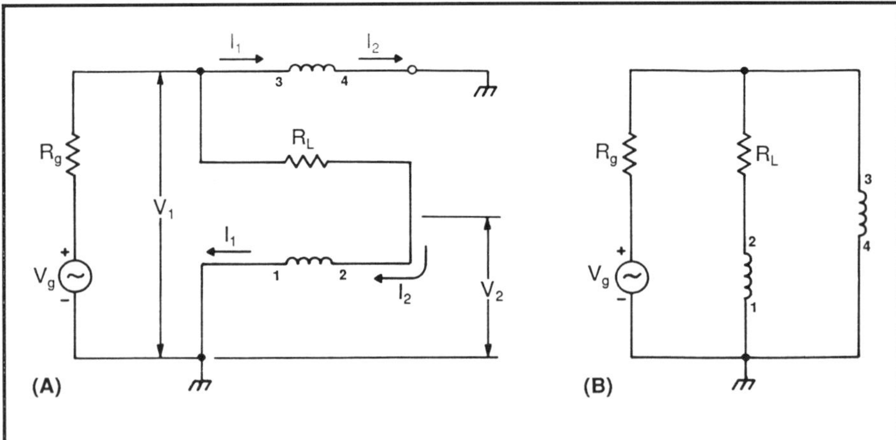

Figure 2-2. Electrical models of the Ruthroff 4:1 balun: (A) high-frequency, (B) low-frequency.

transmission line. Also note, that if the line is electrically one-half wavelength long, the output is *zero.* As a result, Ruthroff's design (which has a built-in cut-off) is sensitive to the transmission line length.

I have recently unearthed another interesting aspect of Ruthroff's design.[2] If the center of the load is grounded, the high-frequency performance is vastly improved. The built-in high-frequency cut-off is eliminated and the balun appears to take on the character of a Guanella balun that sums voltages of equal delays. A closer inspection reveals that the input impedance now consists of two impedances in parallel: one consisting of $R_L/2$, and the other a "flat line" terminated by $R_L/2$. As a result, the currents are not in phase!

Surprisingly, when matching into a folded dipole with an input impedance of 200 ohms, Ruthroff's balun exhibits a high frequency response that is much greater than expected (see later sections). Because the folded dipole has a *virtual* ground at its center, the balun could very well be summing voltages of equal phases.

Sec 2.2.3 Amateur Radio History and Design

Looking back at old issues of amateur radio handbooks (I don't have a complete set), I found that the first presentation on broadband 4:1 baluns appeared in the 1955 edition of *The ARRL Handbook.* The section was called "Coil Baluns." The schematic diagram was that of Guanella's shown in **Figure 2-1A**. What surprised me was that this section appeared to use many of the important words contained in Guanella's article.[3] It mentioned that the choking action of the coiled transmission lines should be great enough to isolate the input from the output at the lowest frequency of interest. It also included the requirement on the characteristic impedance of the coiled transmission lines; namely, that the characteristic impedance should be equal to $R_L/2$, where R_L is the load.

However, the section also included two other statements which are not correct in light of today's design practices. One recommended that the length of the winding in each coil be equal to about a quarter wavelength. The other stated that the principal application is in going from a 300-ohm balanced load to a 75-ohm coaxial line. With magnetic cores, the lengths of the windings are now considerably shorter than a quarter wavelength, and the applications include a host of different impedance levels.

Recent issues of the handbooks now include the 4:1 broadband *coil balun* (along with the same write-up that appeared in the 1955 issue), and one with windings on ferrite cores. They are now called 4:1 air-core current baluns and "just plain" 4:1 current baluns (ferrite cores being assumed). Unfortunately, what is lacking in the description of the 4:1 current balun is information on the importance of the characteristic impedance of the windings and the value of the permeability of the ferrite cores. The literature states that 8 to 10 turns (of No. 14 Formvar™-coated, close-spaced, I guess) on a toroidal core or 10 to 15 turns on a rod are typical values for the HF range. Ferrites with permeabilities from 850 to 2500 are also suggested. Nothing is mentioned regarding the dimensions of the cores.

In essence, there's very little information available today in our handbooks that would help one understand and construct the "popular" *current* balun. Even the choices of recommended ferrites to be used are found wanting. Accurate loss measurements[2] have shown ferrites with permeabilities of 850 to 2500 to be lossy in balun and unun applications. Only when the permeabilities of ferrites are 300 or less, will baluns and ununs exhibit the very high efficiencies of which they are capable.

Even though the 4:1 *voltage* balun has actually had a shorter history than the *current* balun, considerably more construction detail (including an actual photograph) has been available in the amateur radio handbooks. As far as I can tell, the first presentation took place between 1965 and 1968. In looking through succeeding issues (including the 1993 issue), I find the write-up hasn't changed much (if any) over the years. The 4:1 balun from the handbooks and a commercial rod-type are described and compared with my designs in the succeeding sections.

Sec 2.3 4:1 Ruthroff Balun Designs

As mentioned above, the schematic of Ruthroff's balun is shown in **Figure 2-2A**. A pictorial representation is shown in **Figure 2-3**. **Photo 2-A** (on the left) shows my construction of a design close to the one shown in the handbook's photographs. It has 10 close-spaced, bifilar turns of No. 14 Formvar-coated wire on a 2.4-inch OD ferrite toroid with a permeability of 40. **Figure 2-4** shows a plot of the input impedance versus frequency when the 200-ohm load is center-tapped-to-ground (which is close to the actual case when matching into balanced antenna systems). As you can see, when compared to a design that has the proper characteristic impedance of the winding and sufficient choking, the response is very poor. Although this balun has been rated at 1000 watts of RF power from 1.8 through 60 MHz, I would suggest it not be used below 6 MHz for fear of excessive flux in the core (especially when the magnitude of the load is greater than 200 ohms). Also, above 14 MHz, the transformation becomes considerably greater than 4:1.

My design, on the right in **Photo 2-A**, has 14 bifilar turns of No. 14 tinned copper wire on a 2.4-inch OD ferrite toroid with a permeability of 125 or 250. The wires are threaded through No. 13 Teflon™ tubing with a wall thickness of 20 mils. As you can see by its excellent high frequency response in **Figure 2-4**, the characteristic impedance of the bifilar winding must be very close to the ideal value of 100 ohms. **Photo 2-B** shows two different views of my design mounted in a 4 inch long by 3 inch wide by 2.25 inch high Bud CU 234 aluminum box. The balun, which is placed equidistant between the top, bottom, and sides of the enclosure, is securely mounted by soldering its leads

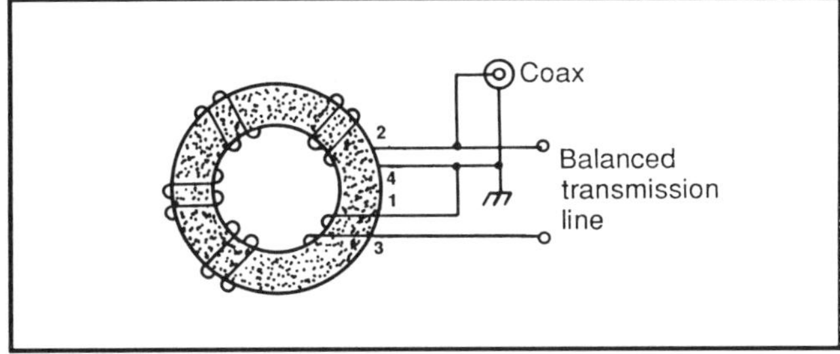

Figure 2-3. A pictorial of the connections for a 4:1 Ruthroff (voltage) balun.

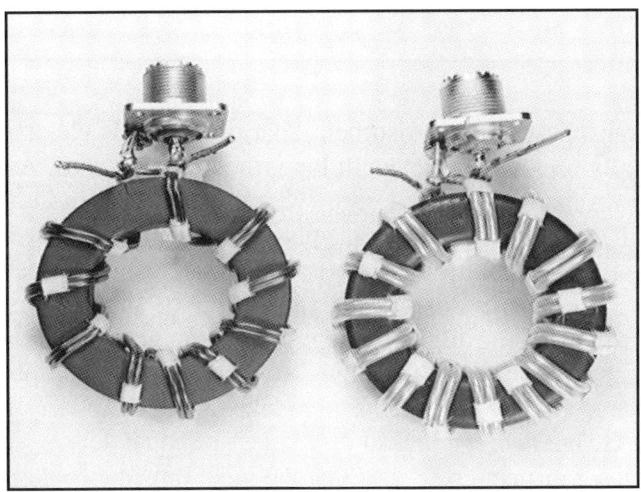

Photo 2-A. Two designs of the 4:1 Ruthroff (voltage) balun. The one on the left is taken from the amateur radio handbook. The one on the right is my improved version.

to the two feedthrough insulators and the SO-239 chassis connector.

It should be mentioned that if the balun is to be used mainly on the lower portion of the HF band (including 160 meters), then the 250 permeability ferrite is recommended. Even though the difference in low frequency response between permeabilities of 125 and 250 doesn't show up in **Figure 2-4**, the 250 permeability would provide an extra safety margin (from flux in the core) at the low frequency end. The trade-off lies in giving up a little in efficiency (about 1 percent) for an increase in the safety margin (a factor of 2) at the low end.

Incidentally, the handbooks also state that the balun can be used between a balanced 300-ohm point and a 75-ohm unbalanced line. Because I suspected this statement, as well, I again measured the input impedances versus frequency of both baluns when terminated in a 300-ohm center-tapped-to-ground load. **Figure 2-5** shows the deterioration that takes place, especial-

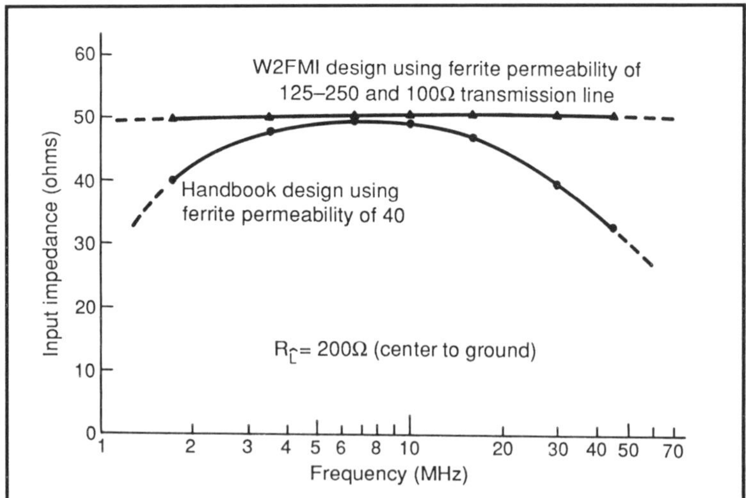

Figure 2-4. The input impedance versus frequency for a 4:1 Ruthroff (voltage) balun design from the amateur radio handbook and one optimized for the 50:200-ohm level. The load is grounded at its center.

Figure 2-5. The input impedance versus frequency for the two Ruthroff baluns of Figure 2-4, but with a 300-ohm load. Note the deterioration of the W2FMI design, which was optimized for the 50:200-ohm level.

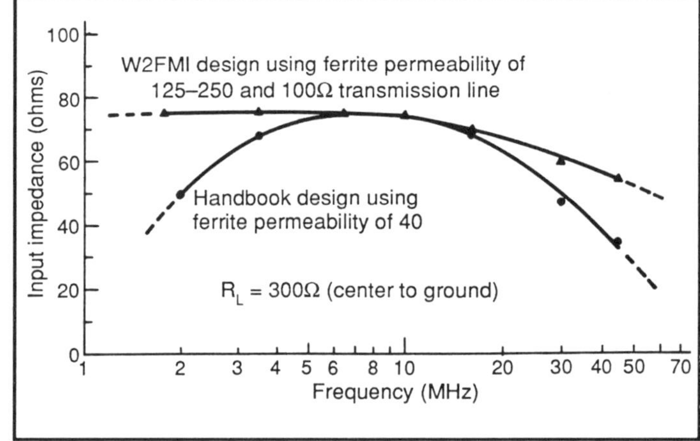

ly at the high end. Even a balun that is well designed for a 50:200 ohm impedance level is not recommended for the 75:300 ohm level. Because the length of the transmission line becomes significant beyond 10 MHz, standing waves then change the impedance ratio due to the mismatch with the balun's transmission line. My design also shows more safety margin at the low end. I'm surprised that these simple measurements weren't made many years ago.

However, the balun shown in the handbook does have one interesting feature. It uses a very low permeability ferrite (40), which has been shown by very accurate insertion loss measurements[2] to yield effi-

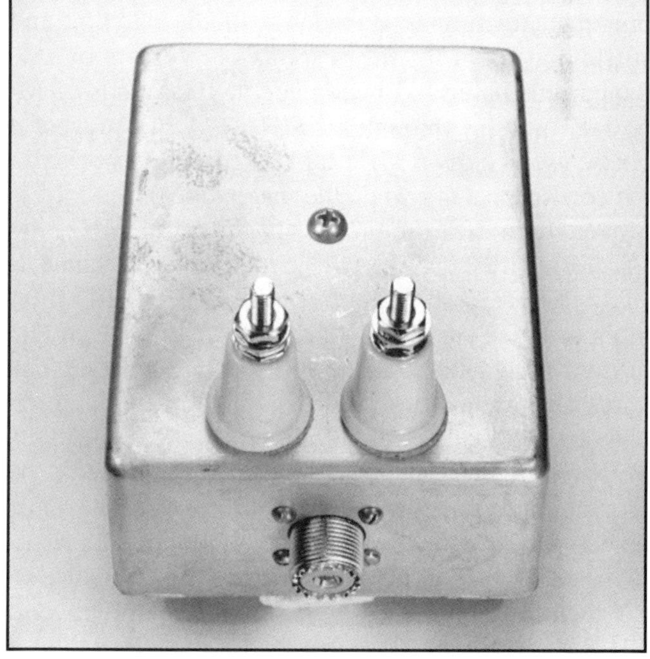

Photo 2-B. Two different views of the optimized version of the 4:1 Ruthroff balun mounted in a 4 inch long by 3 inch wide by 2.25 inch high Bud CU 234 aluminum enclosure.

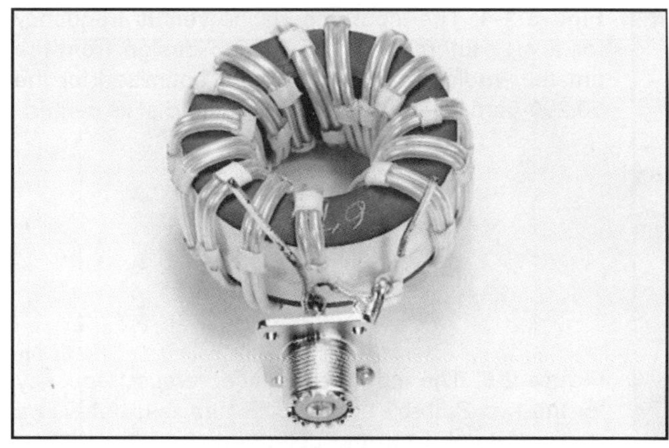

Photo 2-C. Three views of the very high-power 4:1 Ruthroff balun using two low permeability (40) ferrite cores. Dimensions of the aluminum enclosure are: length, 5 inches; width, 3.5 inches; height, 2.25 inches.

ciencies in baluns (and ununs) of *99 percent* at the 50:200 ohm impedance level! This is even a percent or two better than the ferrite with a permeability of 125. Because this ferrite permeability is so low, the major problem lies in obtaining sufficient choking reactance at the lowest frequency of interest, so that only transmission line currents are allowed to flow.

The design chosen (in order to exploit this very high efficiency) is shown **Photo 2-C**. It uses 14 bifilar turns of the same wire, as with my previous balun shown on the right in **Photo 2A**, on two 2.4-inch OD cores (bound together with No. 27 glass tape) with permeabilities of 40. The unmounted view shows how the two cores are bound together by glass tape. The other views attempt to give an example for mounting the balun. The balun is supported by two acrylic end pieces which are, in turn, held fast to the enclosure by a long bolt. The balun is placed equidistant between the top, bottom, and sides of a 5 inch long by 3.5 inch wide by 2.25 inch high aluminum enclosure. A few washers at the point where the bolt comes through enclosure help to position the balun between the top and bottom.

When matching 50 ohms (unbalanced) to 200 ohms (balanced), the response of this balun is practically the same as mine, shown in **Figure 2-4** using a single core. From 1.7 to 30 MHz, it can certainly handle the full legal limit of amateur radio power with an efficiency close to 99 percent. However, if the operation of this balun is restricted to the HF band only (that is, from 3 to 30 MHz), then it could be rated conservatively at 10 kW of peak power and 5 kW of average power. It would be an ideal balun for a log-periodic beam antenna.

Finally, **Photo 2-D** shows three different views of a low power 4:1 Ruthroff (voltage) balun designed to handle the output power of any HF transceiver easily. It has 10 bifilar turns of No. 18 hook-up wire on a 1.5-inch OD ferrite toroid with a permeability of 250. The enclosure is a 2.75 inch long by 2.125 inch wide by 1.625 inch high CU 3000-A minibox.

Section 2.4 4:1 Guanella Balun Designs

Photo 2-E shows two high-power Guanella 4:1 baluns designed to match 50-ohm coaxial cable to loads of 200 ohms. They both use No.14 H Thermaleze wire with a covering of Teflon tubing giving charac-

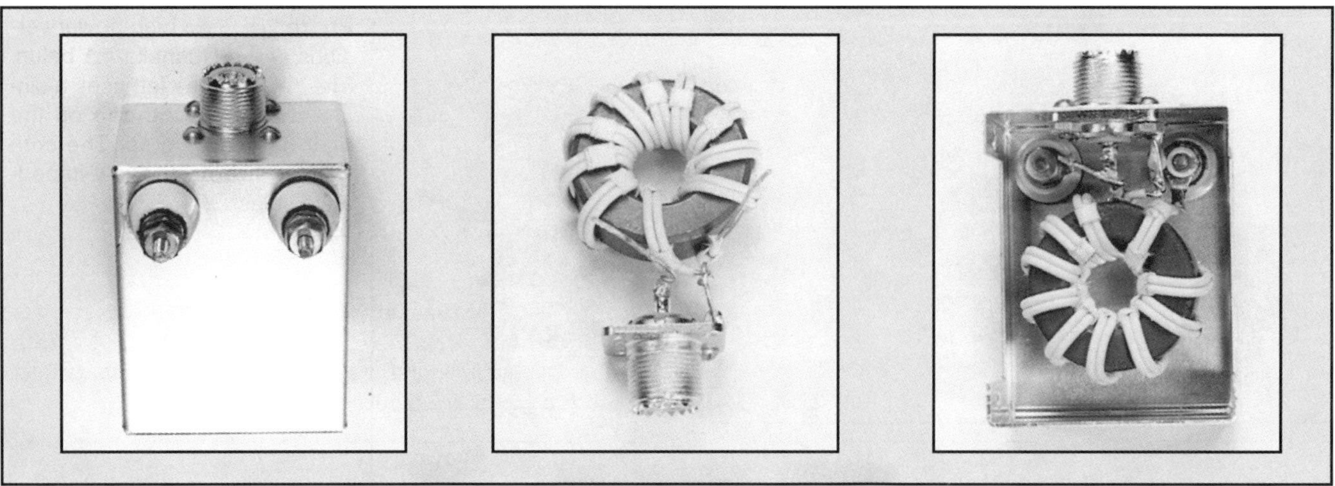

Photo 2-D. Three views of the low-power 4:1 Ruthroff balun designed to handle the output of any HF transceiver. The aluminum enclosure is a 2.75 inches high by 2.125 inches wide by 1.625 inches high CU 3000-A minibox.

teristic impedances very close to 100 ohms (the objective). Their responses are flat from 1.5 MHz to well beyond 30 MHz. Both can easily handle the full legal limit of amateur radio power.

The single-core version (on the left) has 8 bifilar turns on each of its two transmission lines. The dual-core version (on the right) has 16 turns on each core. The wires are clamped together with strips of Scotch No. 27 glass tape placed about every 3/4 inch. The cores are 2.4-inch OD ferrite toroids with a permeability of 250. The connectors are on the low-impedance sides. For ease of connection, the dual-core version has one winding clockwise and the other, counterclockwise. Also, in the dual-core case, the spacing between the two cores (which isn't critical) can be as small as 1/4 inch.

These transformers can also be wound with ordinary No. 14 (solid) house wire. The several samples I tried yielded characteristic impedances close to 100 ohms (and, thus, were acceptable). The major difference lies in the voltage-breakdown capability. Units wound with Teflon-sleeved No. 14 H Thermaleze wire have been reported to withstand 10,000 volts without breakdown! Obviously, this is beyond the capability of ordinary house-wire.

Photo 2-F shows a Guanella (current) 4:1 balun mounted in a 5 inch long by 4 inch wide by 3 inch high CU 3005-A minibox. It has 14 bifilar turns of No. 14 H Thermaleze wire on each of the two 2.4-inch OD ferrite toroids with permeabilities of 250. Each wire is covered with Teflon tubing, resulting in a characteristic impedance close to 100 ohms (the optimum). The windings also employ a crossover after the seventh turn, as shown in **Figure 2-6**. For ease of connection, one toroid is wound clockwise and the other, counterclockwise. The spacing between the toroids can be between 1/4 and 1/2 inch.

When matching 50-ohm cable to a balanced load of 200 ohms, the transformation ratio is constant (within 2 percent) from 1.5 to 45 MHz. This balun can also handle the legal limit of amateur radio power. It would probably perform satisfactorily if wound with ordinary No. 14 house wire (solid), or with Teflon-covered No. 14 tinned wire. However, the design in **Photo 2-F** has withstood peak pulses of 10,000 volts! Considering this fact, it might be worthwhile to take the extra effort and use Teflon-covered No. 14 H Thermaleze wire.

A single-core version using two coiled transmission lines on a single core looks interesting and should be investigated. It results in balanced currents and unbalanced voltages. I would use two coiled transmission lines with 7 bifilar turns of the same wire on the same core as above.

Photo 2-G shows two low-power Guanella 4:1 baluns designed to match 50-ohm coaxial cable to loads of 200 ohms. They both use No. 20 hook-up wire (solid) giving a characteristic impedance very close to the objective of 100 ohms. Their responses are flat from 1.5 MHz to well beyond 50 MHz. They are conservatively rated at 150 watts of continuous power and 300 watts of peak power. They have been exposed to 500 watts of continuous power (in a matched condition) for a considerable length of time with virtually no rise in temperature.

The single-core version (on the left) has 7 bifilar turns on each of its two transmission lines, while the

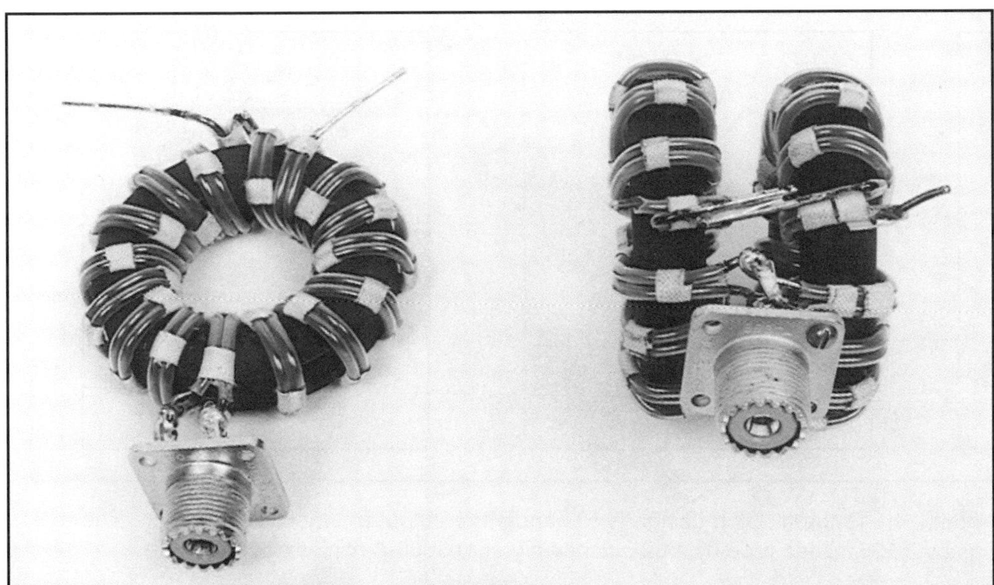

Photo 2-E. Two high-power versions of the Guanella 4:1 balun. The balun on the left uses a single core while the one on the right uses two cores. The connectors are on the low-impedance sides.

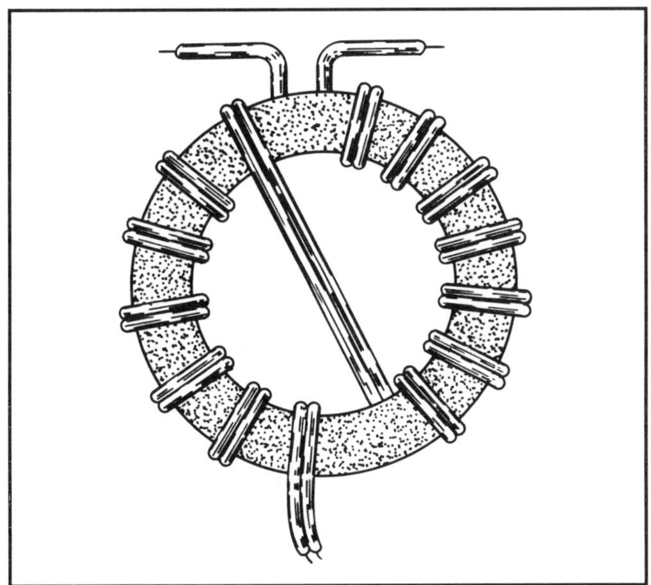

Figure 2-6. Construction of a crossover placing input and output connections on opposite sides of the toroid.

dual-core version (on the right) has 14 turns on each core. The wires are clamped together about every 1/2 inch with strips of Scotch No. 27 glass tape. The cores are 1.25-inch OD ferrite toroids with a permeability of 250. The connectors are on the low-impedance sides. As above, the dual-core version has one winding clockwise and the other, counterclockwise.

Photo 2-H is a step-down version of the Guanella 4:1 balun, and uses two ferrite rod-cores 3/8 inch in diameter and 3.5 inches in length. Core permeabilities are 125. It uses the schematic of **Figure 2-1A,** but with the generator (which is grounded) on the right side and the load (ungrounded) on the left side. This 4:1 balun is designed to match 50-ohm coaxial cable (on the right side) to a balanced load of 12.5 ohms. Each rod has 13.5 bifilar turns of No. 14 H Thermaleze wire. Again, for ease of connection, one rod is wound clockwise and the other, counterclockwise. The response is flat from 1.5 MHz to well over 30 MHz. This balun is fully capable of handling the legal limit of amateur radio power. The connector is on the high-impedance (50 ohms) side. Beaded versions of Guanella's step-down 4:1 balun also look very promising for operations on the VHF and UHF bands. The technique requires minimizing the parasitics in the interconnections. Examples will be shown later.

It should be mentioned again that the three dual-core baluns above also make excellent broadband ununs. They only sacrifice a little in low-frequency response. However because of their conservative designs, they can still handle the 160-meter band.

Sec 2.5 Comparisons with Other Baluns

After completing the study on 4:1 baluns, I thought it would be interesting to characterize other baluns that are commercially available, or that have been recently described in the amateur radio literature. My findings are as follows:

The 4:1 Rod-type Ruthroff Balun

Photo 2-I shows a photograph of the typical rod-type 4:1 balun, which has been practically the only one available over the past two to three decades. The balun in the photograph is the **HI-Q Balun.** It is the

Photo 2-F. A 4:1 Guanella balun using the crossover connection of Figure 2-6 and mounted in a 5 inch long by 4 inch wide by 3 inch high CU 3005-A minibox.

Ruthroff design (now called a *voltage* balun[6]) with 10 bifilar turns of No. 14 wire on a 1/2-inch diameter ferrite rod 2 inches in length. In terminating this balun with 200 ohms, the useful range was found to be from 7 to 15 MHz. Below 7 MHz, the input impedance showed a considerable inductive component—indicating autotransformer action and flux in the core (which could be harmful). Above 15 MHz, the transformation ratio increased and became complex. The optimum impedance level was found when matching 100 ohms to 25 ohms (indicating a characteristic impedance of the windings of only 50 ohms). The useful frequency range at this impedance level increased from 3.5 MHz to 30 MHz.

When matching 50-ohm coaxial cable to a 20-meter folded dipole at a height of 0.17 wavelengths (result-

Photo 2-G. Two low-power versions of the Guanella 4:1 balun. The balun on the left uses a single core while the one on the right uses two cores. The connectors are on the low-impedance sides.

Photo 2-H. A dual rod-core 4:1 Guanella step-down balun designed match 50-ohm cable to a balanced load of 12.5 ohms. The connector is on the 50-ohm unbalanced side. →

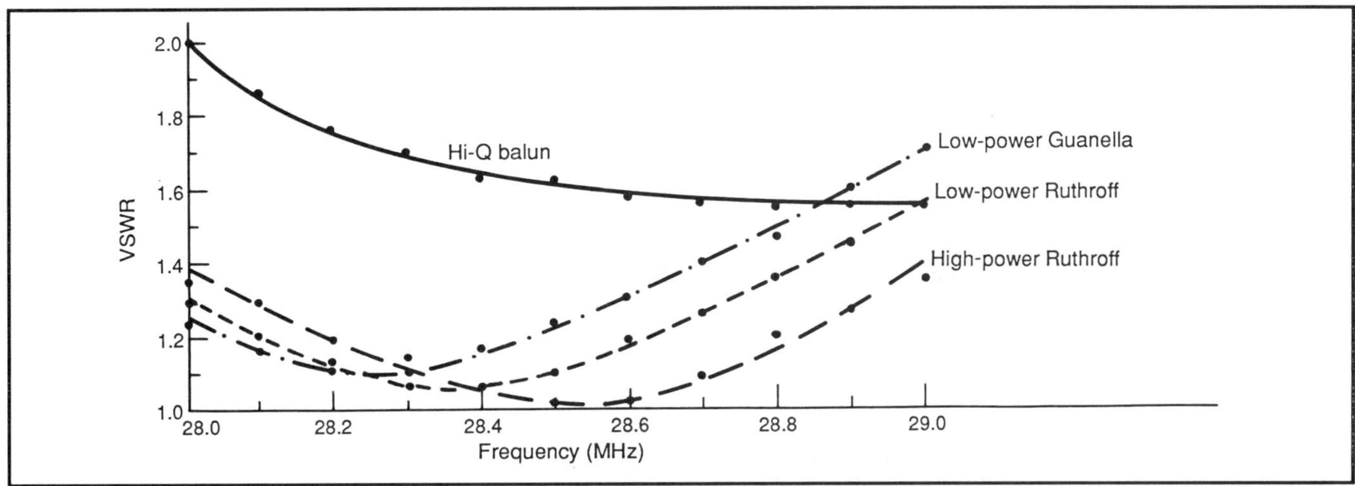

Figure 2-7. Plots of VSWR curves on 10 meters for four different 4:1 baluns. The high-power Ruthroff balun is actually McCoy's design with a 100 ohm characteristic impedance winding (and is described in the next chapter). The Hi-Q balun is shown in Photo 2-I. The comparisons show the importance of having the optimum value of the characteristic impedance of the windings and that the high-power balun with its much longer transmission line indicates a Guanella-type operation.

ing in a resonant input impedance of 200 ohms), the VSWR curve was indistinguishable from that of the best Guanella 4:1 (current) baluns.[6] This balun also presented no difficulty in handling the full power limit. However, on 10 meters, the difference due to a very low characteristic impedance of the coiled transmission line became evident. **Figure 2-7** shows the poor VSWR curve of the rod-type balun when compared to other Guanella and Ruthroff baluns with characteristic impedances close to the optimum value of 100 ohms. The rod-type balun with all of its inadequacies, which include voltage-breakdown, is certainly *not* recommended.

The high-power Ruthroff balun is close to McCoy's design.[17] It uses 11 bifilar turns of No. 14 H Thermaleze wire on a stack of three T200-2 cores. The wires are also covered with 20-mil wall Teflon tubing yielding a characteristic impedance close to 100 ohms. I differ with McCoy's design because I feel the characteristic impedance of his transmission line could be closer to 50 ohms. His balun and others that use powdered-iron cores will be described further in the next chapter on baluns for antenna tuners.

The low-power units in **Figure 2-7** used 14 bifilar turns of No. 18 hook-up wire on 1.25 inch OD ferrite toroids with a permeability of 250. The transmission lines of the low-power baluns were just under 20 inches, while those of the high-power balun were 50 inches in length.

As can be seen in **Figure 2-7**, there are **very** small differences between the VSWR curves of the two low-power baluns and the high-power Ruthroff (voltage) balun. The differences could very well be attributed to the small variations in the characteristic impedances of the windings. Very likely, the most important information gleaned from **Figure 2-7** is that, when feeding a balanced dipole with a virtual ground-plain bisecting it, the Ruthroff balun takes on the character of a two-core Guanella balun. In other words, the Ruthroff balun loses its built-in high-frequency cut-off!

4:1 Current Baluns

I also characterized several so-called *current* baluns[6] that recently appeared on the market. These are my findings:

a) They are the dual-core (toroids) version of the Guanella balun, which sums voltages of equal delays.

b) The electrical performances of these baluns are vastly superior to the rod-type balun described earlier.

c) These baluns should meet their electrical and power-rating specifications.

d) My only criticism is that they could have more of a safety margin at the low-frequency end, where excessive core flux (due to higher than expected impedances) could take place. More inductance in the windings is recommended.

The Beaded-coax 4:1 Balun

A recent design in an amateur radio journal[5] advocated using beaded coaxial cable (of 100 ohms) in a 4:1 Guanella design. Various claims were advanced for this "new" approach. I constructed one of these baluns using No. 14 wire with Teflon sleeving result-

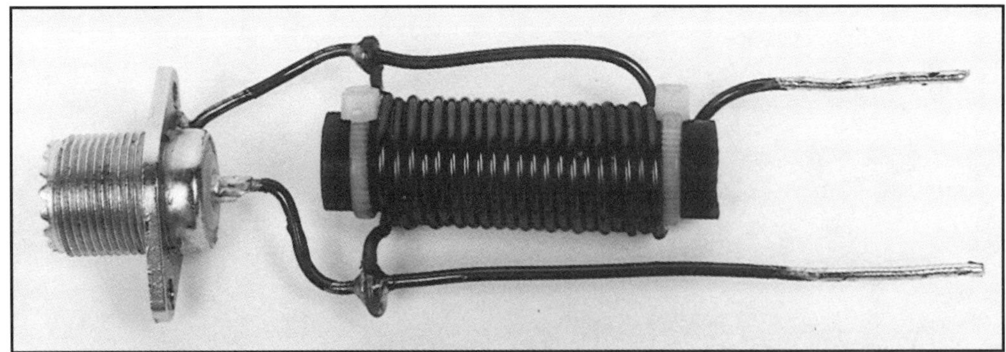

Photo 2-I. A typical 4:1 rod-type Ruthroff (voltage) balun (Hi-Q).

ing in the required 100-ohm characteristic impedance. Here are my findings:

a) The balun had excellent margins at both the high- and low-frequency ends. The performance of this balun verified the analysis (expressed earlier) with the high- and low-frequency models and the subsequent voltage gradients. In fact, the high-frequency performance exceeded the capability of my simple test equipment.[2]

b) The major disadvantage is in efficiency. Because high-permeability (2500) beads are required in order to obtain the required choking reactance in the HF band, this balun had considerably more loss than coiled-type baluns using low-permeability (less than 300) ferrite toroids.[2] A soak-test[2] (transformers connected back-to-back and about 500 watts applied into a dummy load) with the dual-core low-power unit in **Photo 2-G** showed that the smaller balun ran considerably cooler! The beaded transmission line technique is recommended mainly for baluns (and ununs) operating at low impedance levels or on the higher-frequency bands.

Sec 2.6 Summary

Unlike the 1:1 balun, the 4:1 balun matching 50 ohms unbalanced to 200 ohms balanced has had no real standard for comparison in the amateur radio literature. As **Chapter 1** showed, Reisert's balun in his 1978 article[7] had all of the attributes of a good 1:1 design. Therefore, he set a legitimate standard for others to follow or even attempt to exceed. Also shown were some of my variations in his design for increased efficiency and ease of construction.

However, **Chapter 2** has illustrated that the designs in the amateur literature (particularly the handbooks) are found lacking in bandwidth, or efficiency, or both. Even the 4:1 baluns on the commercial market can be improved. This is *especially* true of the rod-type balun that has been available for more than three decades!

In the process of investigating the 4:1 balun for my series of articles in *CQ* and *Communications Quarterly*, I have arrived at some designs that could provide the beginnings of standards for this device. They are included in this chapter and are:

1. For *balanced* applications like matching 50-ohm cable to the 200-ohm balanced input impedance of folded-dipole or log-periodic antennas, I recommend the single-core, Ruthroff design of **Photo 2-A**. It is capable of handling the full legal limit of amateur radio power. For a higher-power capability and a little less bandwidth, I recommend the Ruthroff design shown in **Photo 2-C**. These designs are presently called *voltage* baluns.[6]

2. For unbalanced applications like the OCFD (off-center-fed-dipole), or a dipole that could be unbalanced by surrounding structures or by construction errors, I recommend the two-core Guanella design of **Photo 2-E** or **Photo 2-F**. It is a much more flexible unit that can operate successfully as a balun when the load is grounded at its center (**Figure 2-1A**), or as an unun when the load is grounded at the bottom. Although not shown, it can even be grounded at the top yielding a 4:1 phase-inverter.

3. For low-power 4:1 baluns, there really have been no designs in the literature for comparisons. Therefore, by default, the designs in this chapter are suggested as standards. Applications of the single-core Ruthroff balun and the two-core Guanella balun are the same as their higher-power counterparts.

I am sure there are some that don't agree with the recommendations proposed above. The Guanella (current) balun now appears to be the only balun of choice. Manufacturers even stress in their ads that their designs are "current" baluns. The question is: Why use a two-core Guanella (current) balun when a single-core Ruthroff (voltage) balun will do? They both have the same power ratings! For those that disagree with my views, designs or recommendations, I encourage them to (as the popular TV commercial says) *put it in writing*. Then we will all benefit from the new information.

Chapter 3

Baluns for Antenna Tuners

Sec 3.1 Introduction

The 4:1 balun, matching 50 ohms (unbalanced) to 200 ohms (balanced), has found its most popular use in antenna tuners. Because the balun rarely sees a resistive load of 200 ohms in this application, the primary objective is to take the balanced impedance (with respect to ground) of the input to an open-wire (or twin-lead) feedline and transform it into an unbalanced impedance which has one side grounded and can be transformed into 50 ohms by an L-C matching network. This was well described in two recent *CQ* articles by McCoy.[17,18]

As **Chapter 2** has shown, there are two different forms of the 4:1 balun. One uses two transmission lines wound on separate cores (or threaded through ferrite beads in some cases) and connected in parallel at the 50-ohm side and in series at the 200-ohm side. This design was first presented by Guanella in 1944,[3] and is presently called a *current* balun.[6] The other design, using a single transmission line wound around a core and connected in a phase-inverter configuration, was introduced by Ruthroff in 1959.[9] This design, which has recently been called a *voltage* balun,[6] is now perceived as the inferior design.

This chapter presents another view of the 4:1 balun. It not only includes the optimum design considerations for antenna tuner use, but also the design parameters for multiband antenna systems using center-fed dipoles with open-wire or twin-lead feedlines. Also included are my views on the G5RV antenna, the tuner using a 1:1 balun before the L-C matching network, and the special cases of Ruthroff's 4:1 design matching into a load that is actually or virtually grounded at its center.

Sec 3.2 The Two Forms of Antenna Tuners

Three of the more common items in current amateur radio jargon are: VSWR, antenna tuners, and multiband antennas (especially the G5RV). These have appeared upon the scene because of the ease at which bands can now be changed and the narrow limits in the range of matching impedances with modern rigs.

The concept of using a wire antenna on many different bands isn't new. Designs have been around for more than five decades. In fact, satisfactory circuits have also been available which couple transmitters to balanced lines that present loads different than the transmitter output impedance. These were known as series and parallel-tuned circuits.[19] The transforming of a balanced impedance to an unbalanced impedance was accomplished by the isolation provided by magnetic coupling. Energy was transmitted from one cir-

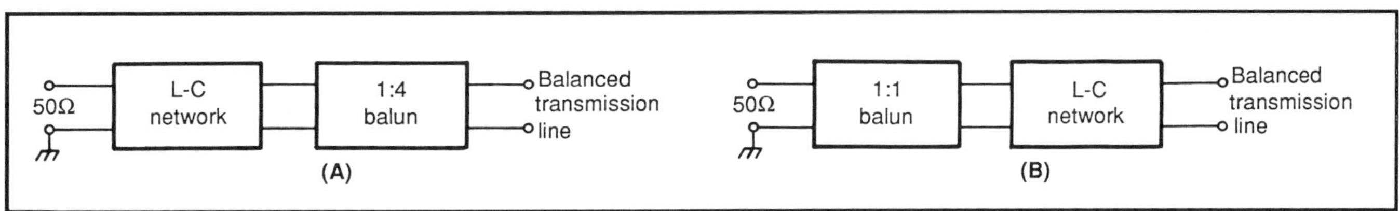

Figure 3-1. The two basic forms of the Transmatch (antenna tuner): (A) The more popular design using an unbalanced L-C network and a 1:4 balun; (B) a 1:1 balun and a balanced L-C network.

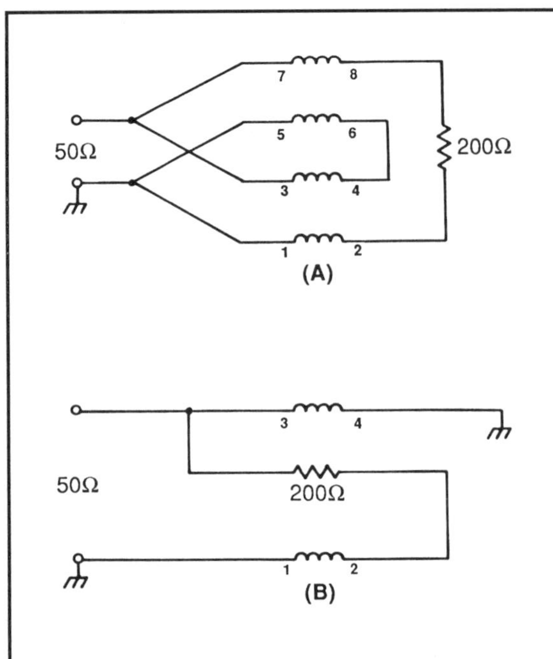

Figure 3-2. The two basic forms of the 1:4 balun: (A) The Guanella (current) balun; (B) the Ruthroff (voltage) balun.

Figure 3-3. The frequency response of a 4:1 Ruthroff (voltage) balun with the load floating and with its center grounded. In the grounded case, the high-frequency response is similar to that of a 4:1 Guanella (current) balun.

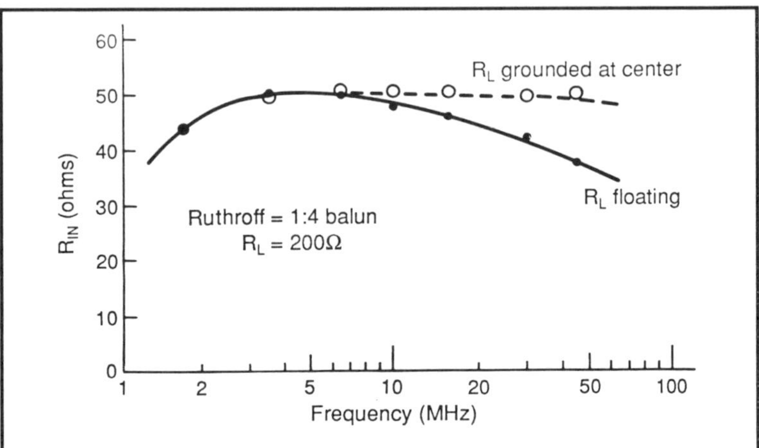

cuit to the other by either having two coils in close proximity or by "link" coupling. However, these methods of coupling have fallen by the wayside together with rock-bound rigs, plug-in coils, separate receivers and transmitters, and (sad to say) the exciting flashing of mercury-vapor rectifiers.

Today, the *transmatch* is most often used to convert the reactive/resistive load presented by an antenna system to a nonreactive, grounded 50-ohm load. The transmatch is also commonly known as an antenna tuner. The isolation role, that of converting a balanced impedance to an unbalanced one, is now provided for by the balun transformer.

There are two basic forms of the transmatch, and they are shown in **Figure 3-1**. **Figure 3-1A**, which shows a 4:1 balun between the L-C network and the balanced transmission line, has been the most popular. In some designs, a 1:1 balun has been used. In either case, this form of an antenna tuner places the burden on the balun, not on the L-C network. Depending upon the dimensions of the antenna and open-wire (or twin-lead) transmission line, the balun can see very high impedances that may be harmful. In turn, the L-C networks are simple because of their unbalanced nature. Among the most popular networks are L, pi, T, Ultimate, and SPC types.[20–23] An added advantage to this approach is that the balun can be placed outside the operating area and connected to the L-C network by a coaxial cable.[17,18]

On the other hand, **Figure 3-1B** takes the complexity out of the balun and places it on the L-C network. With a balanced network, the 1:1 balun should see a lower voltage drop along the length of its transmission line (and, hence, less loss[2]) because its load is always close to 50 ohms. Additionally, the choking requirements of a 1:1 balun are considerably less than that of the 4:1 balun in **Figure 3-1A**.

Roehm has addressed the problems related to this form of transmatch in a recent article.[16] In fact, he suggests a design using an unbalanced T network and a 1:1 beaded-coax balun. Although a balanced L-C network is inherently more complex and costly, it would be interesting to see its comparison with Roehm's unbalanced design. Additionally, a comparison with a 1:1 balun using 50-ohm twin-lead or coaxial cable wound around a ferrite toroid with a permeability of less than 300 would also be useful. Because the 1:1 beaded-coax (choke-type) balun requires ferrite beads with permeabilities considerably greater than 300, it has more loss.[2]

In any event, the basic form of the transmatch using the design in **Figure 3-1B** looks promising and merits further investigation.

Sec 3.3 Another View of the 4:1 Balun

This section presents my views and the results of my work on a 4:1 balun designed for use in the very popular "antenna tuner" shown in **Figure 3-1A**. Because

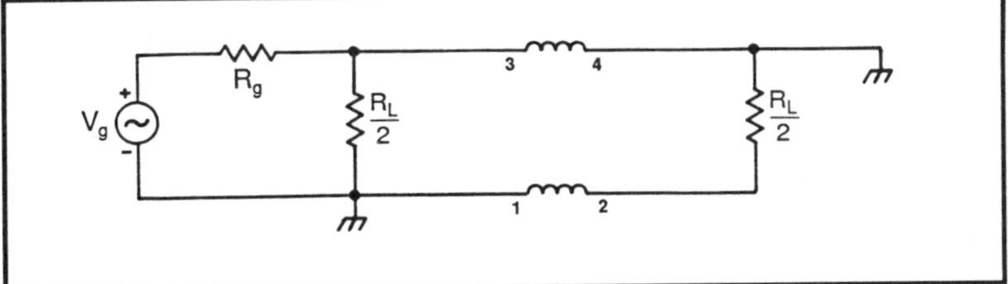

Figure 3-4. Suggested model of the 4:1 Ruthroff (voltage) balun when the load, R_L, is grounded at its center.

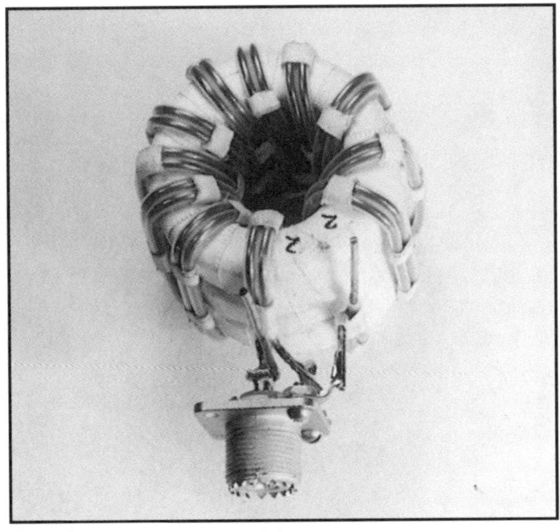

Photo 3-A. The high-power Ruthroff 4:1 balun used in the comparison with other baluns (see Figure 3-7, and Figure 2-7 of Chapter 2). Except for some difference in the characteristic impedance of the bifilar winding, it is essentially the McCoy 4:1 balun.

the balun may be exposed to harmful high voltage conditions in this application, the efficiency and ruggedness of the core materials are important considerations. Experiments have shown that losses in baluns are related to the impedance level[2] (and, hence, voltage level), and the permeability of the materials. Therefore, the losses are of a dielectric-type and not of a current-type, as in conventional transformers. Moreover, it is well known that powdered-iron is a more rugged and linear material than ferrite. A very important question to ask is which form of the 4:1 balun should be used in this application—Guanella's or Ruthroff's? My conclusions may surprise many readers.

As was mentioned at the beginning of this chapter and above, there are two basic forms of the 4:1 balun. They are shown in **Figure 3-2A** and **B**. **Figure 3-2A** is Guanella's approach. It uses two coiled transmission lines (on separate cores) connected in parallel on the 50-ohm side and in series on the 200-ohm side. This has recently been called a *current* balun.[6] In order to have "flat" transmission lines and obtain the highest frequency response, the characteristic impedance of the coiled transmission lines should be equal to the loads they see—namely, $1/2R_L$ and, in this case, 100 ohms.

As Guanella said in his classic paper,[3] this balun is literally "frequency independent." At the low frequency end, the reactance of the coiled transmission line should be much greater than 100 ohms (in this case) in order to assure that the energy is transmitted from input to output by an efficient transmission line mode. Beaded transmission lines aren't recommended for use on the HF band at these impedance levels because of excessive dielectric loss.

Figure 3-2B shows Ruthroff's approach, which uses a single transmission line connected in the phase-inverter configuration (see **Chapter 1**). By grounding terminal 4, a voltage drop of $-V_1$ appears across the length of the transmission line. As a result, terminal 3 is at $+V_1$ and terminal 2 is at $-V_1$—a 4:1 transformation ratio.

It is when the load is actually grounded at its center, that we see a very interesting feature of this approach. A simple impedance measurement will show that the high frequency response is *vastly* improved and that the Ruthroff (voltage) balun appears to take on the character of a Guanella (current) balun! **Figure 3-3** illustrates the measurements of the input impedance of a Ruthroff 4:1 balun with the load floating, and when it is grounded at its center. These measurements (with a simple resistive bridge) were made on my design using a powdered-iron core, which will be described later.

A model for the Ruthroff 4:1 balun, when the load is grounded at its center, is provided in **Figure 3-4**. If the characteristic impedance of the transmission line in the 4:1 balun is 100 ohms (the optimum value if the load is 200 ohms), then the generator sees $R_L/2$ in parallel with a $R_L/2$ from a "flat" line. As a result, the generator sees its match of 50 ohms—even though the currents in the loads are not in phase!

However, when the balun is connected to a center-fed, folded dipole, or log-periodic beam antenna with 200-ohm input impedances, the virtual ground-plane bisecting the antennas presents an interesting case. Because there is no metallic connection to the center of the load, the currents in both halves of the antennas are in phase. This is unlike the situation in **Figure 3-4**, where the current in the load on the right is delayed

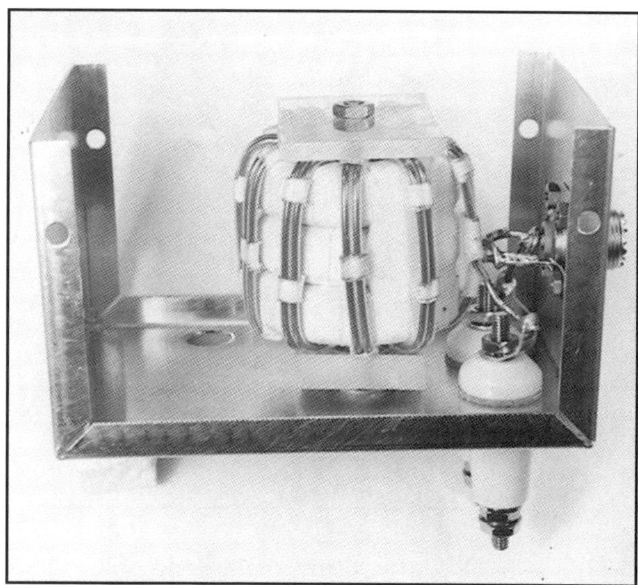

Photo 3-B. The balun of Photo 3-A mounted in a 5 inches long by 4 inches wide by 3 inches high minibox.

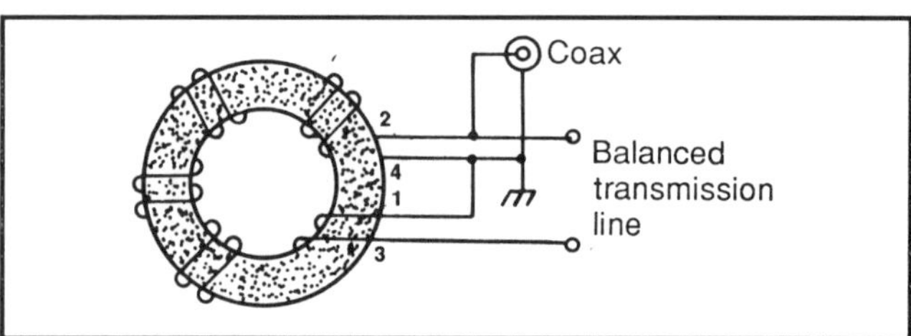

Figure 3-5. A pictorial of the connections for a 4:1 Ruthroff (voltage) balun.

compared to the current in the load on the left. As is evident in **Figure 2-7** of **Chapter 2**, the VSWR curve for the high-power Ruthroff balun has practically the same shape as those of the two lower-power units—even though its transmission line is more than two and one half times longer.

Because the high-frequency response of a Ruthroff 4:1 balun with a floating load is highly dependent upon the length of the transmission line, **Figure 2-7** suggests that, in the virtual ground case, the balun acts as a Guanella balun which sums voltages of equal phases. It also suggests that the effective electrical length of the transmission line is one-half of its actual length. I am quite sure that this model of Ruthroff's balun was not proposed by him or by others that followed. However, it should be remembered that this condition only exists in the *balanced* case. With an unbalanced load, the balun should introduce a reactive component that will limit the high frequency response. In the unbalanced case, the Guanella balun with two cores is the balun of choice.

Sec 3.4 Some "Hardy" 4:1 Designs

Photos 3-A and **3-B** show two views of a Ruthroff 4:1 balun much like McCoy's design, which has been used in his highly popular transmatch.[22] **Figure 3-5** shows a pictorial representation of the connections. It has 11 bifilar turns of No. 14 H Thermaleze wire on three (stacked) T200-2 cores. The cores are pow-

Photo 3-C. The various baluns used in the study of 4:1 baluns for antenna tuners.

dered-iron material with a permeability of 10. The OD is 2 inches. The wires are also covered with a 15-mil wall Teflon™ tubing, yielding a characteristic impedance close to the optimum value of 100 ohms.

I differ with McCoy's design here, because the characteristic impedance of his balun could be closer to 50 ohms. **Photo 3-B** shows the balun mounted in a 5 inch long by 4 inch wide by 3 inch high minibox. Because McCoy didn't use a thickly insulated wire, he wound a layer of Scotch No. 27 glass tape on each toroid before stacking. This was followed by another layer in the stacking process. With Teflon sleeving over the wire, the extra insulation provided by the glass tape could be dispensed with.

In order to improve the low-frequency response of McCoy's balun, I made a study of higher permeability powdered-iron cores. **Photo 3-C** shows the various baluns used in the study. The object of this study was to determine the best core material for a 4:1 balun to be used in antenna tuners where they can be exposed to high impedances (and, hence, hostile environments). I knew, as the result of very accurate insertion loss measurements[2] that loss with ferrite materials was related to the voltage drop along the length of the transmission line and to the value of the permeability. Permeabilities of 40 (No. 67 ferrite) exhibited the lowest loss. The results taken on a single powdered-iron material—No. 2 material with a permeability of 10—also showed the very same low loss. Because powdered-iron material has been known to be more rugged and linear than ferrite material, this suggested that other powdered-irons with greater permeabilities should also be investigated.

I investigated four other powdered-irons with permeabilities of 20, 25, 35, and 75. Their designations were Nos. 1, 15, 3, and 26, respectively. Comparisons were made on input impedances (with the outputs terminated in 200 ohms) and temperature rises (when handling 500 watts of power). The power test showed, convincingly, that No. 26 material was not to be used, because it showed a definite rise in temperature while the other three didn't. However, all four materials showed a definite lower input impedance than the No. 2 material, which has a permeability of 10. As expected, the higher the permeability, the larger the difference with No. 2 material. Although an input impedance measurement does provide some indication of loss because it appears as a shunting path to ground, a very accurate insertion loss measurement would provide a more precise indication of the trade-off that can be made in efficiency for low frequency response.

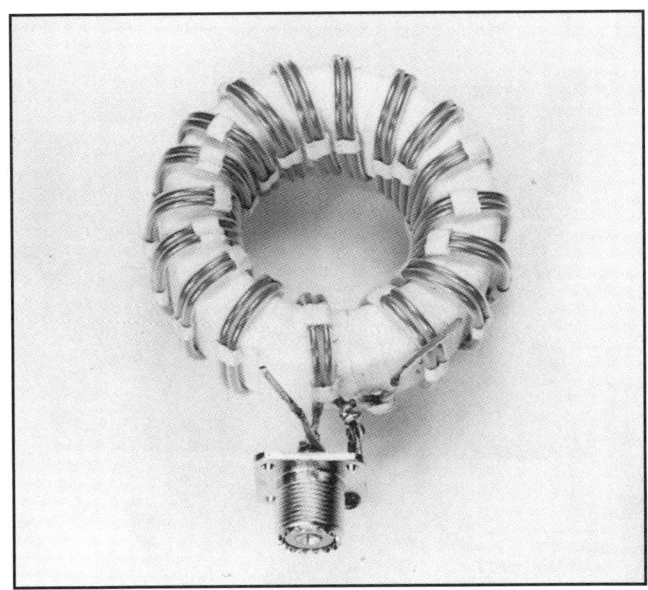

Photo 3-D. An improved 4:1 Ruthroff design for antenna tuners.

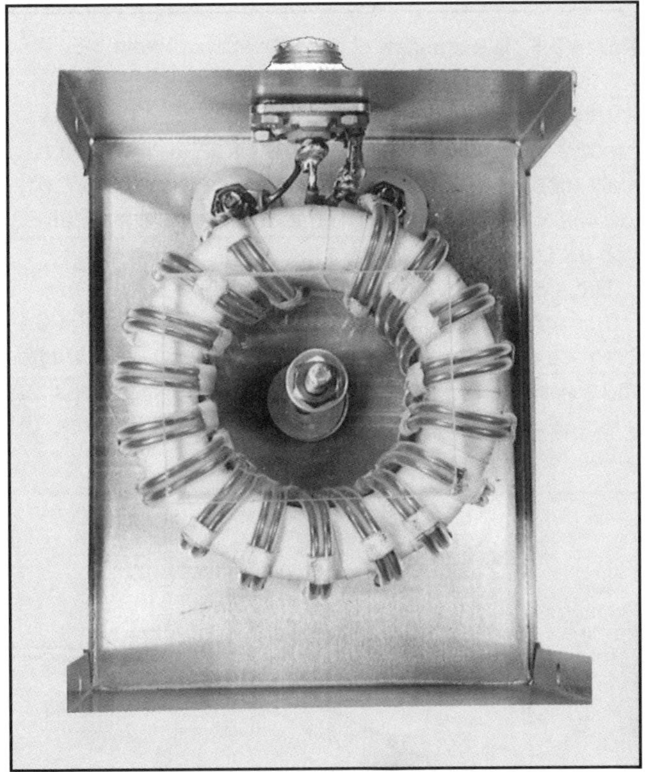

Photo 3-E. The improved 4:1 Ruthroff balun mounted in a 5 inches long by 4 inches wide by 3 inches high minibox.

Because my simple loss measurements indicated that the higher permeability powdered-irons had more loss than the No. 2 material, I decided to design a 4:1 Ruthroff balun using this material—but with a larger core and more turns than the McCoy[17,18] balun.

Photo 3-F. The top view of the 4:1 balun of Photo 3-E.

Although McCoy's design has enjoyed considerable success over the years, I felt that a larger inductive reactance was desirable in order to assure better performance on the lower frequency bands (particularly 160 meters).

The specific design is shown in **Photo 3-D**. It has 17 bifilar turns of No. 14 H Thermaleze wire on a T300A-2 powdered-iron core, which has an OD of 3 inches and a permeability of 10. With this number of turns and a larger cross section than the three T200-2 cores, the low frequency response improved by a factor of about two over the McCoy balun, which has 10 to 12 turns on a stack of three T200-2 cores. Furthermore, the wires are also covered with 15-mil wall Teflon tubing, resulting in a characteristic impedance of 100 ohms (the objective). This well-insulated transmission line has been reported to handle 10,000 volts without breakdown. **Figure 3-3** illustrates the performance of this balun (under a matched condition) when the load is floating and when it is center-tapped-to-ground. **Photo 3-E** shows the balun mounted in a minibox 5 inches long by 4 inches wide by 3 inches high. **Photo 3-F** shows the top view of the mounted balun.

Because the balun with the larger core and more turns showed an improvement by a factor of two in the low-frequency response over the McCoy balun, I constructed an even larger design. This is shown in **Photo 3-G**. It has 21 turns of the same wire on a T400A-2 core with an OD of 4 inches and a permeability of 10. Even though **Photo 3-G** shows the toroid wrapped with Scotch No. 27 glass tape, as mentioned earlier, this extra insulation isn't required because the wires are covered with Teflon sleeving.

Figure 3-6 shows the comparisons in the input impedances versus frequency of these three "hardy" baluns when they are terminated in 200-ohm loads grounded at their centers. As can be seen, the low-frequency response of the balun with the most turns and largest core is the best. Even though the balun with the 3-inch core has a poorer low-frequency response, it is an improvement over the McCoy balun and should find considerable use in transmatches. **Figure 3-6** also shows that the high frequency responses of

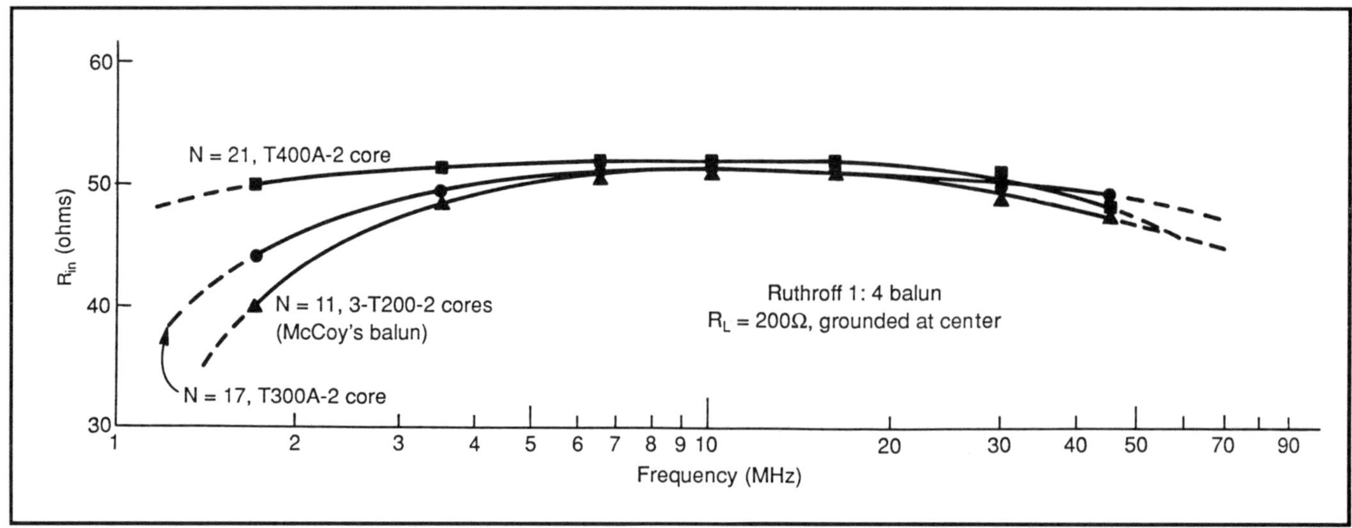

Figure 3-6. A comparison of the performance of the McCoy 4:1 balun with baluns having larger cores and more turns.

these baluns with the loads grounded at their centers are remarkably similar. This is especially interesting, as the length of the transmission line on the larger balun (with the T400A-2 core) is more than twice as long as the other two (115 inches compared to 50 and 55 inches). If it is correct that my model of a Ruthroff (voltage) balun feeding a balanced antenna or transmission line takes on the character of a Guanella (current) balun because of the virtual ground, then this large balun could have many applications.

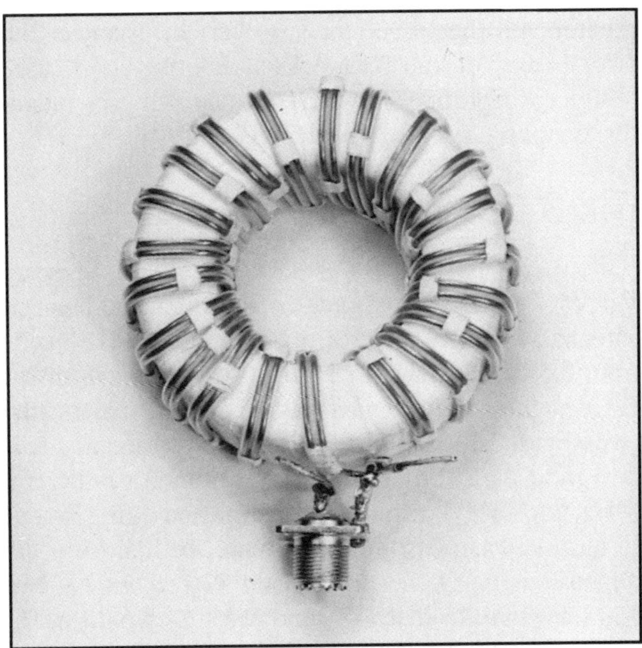

Photo 3-G. A large 4:1 Ruthroff design using a 4-inch OD powdered-iron core and 21 bifilar turns.

Sec 3.5 Multiband Dipoles

Antenna tuners have been known to work well for some radio amateurs and not for others. This is due to the differences in the dimensions of their antenna systems. With high impedances seen by the 4:1 baluns in the antenna tuners, the baluns not only fail to provide a good balanced-to-unbalanced conversion, but they can also be damaged by excessive heating. The high current, low impedance condition seen by the balun isn't a problem. Therefore, the object in multiband antenna design is to provide the most favorable impedances for the baluns, especially on the three lowest frequency bands—40, 80, and 160 meters. Usually on the higher frequency bands, the impedances seen by the baluns are not as high and the balun's choking reactances are greater (assuring balanced-to-unbalanced conversion). This section discusses three cases of multiband center-fed dipole designs. They are 1) the "worst case" design, 2) a smaller design—the G5RV, and 3) a larger design. Others may have better designs for multiband operation, but it's clear that the "worst case" design ought to be avoided. **Figure 3-7** shows the symbols for the dimensions of the center-fed dipoles.

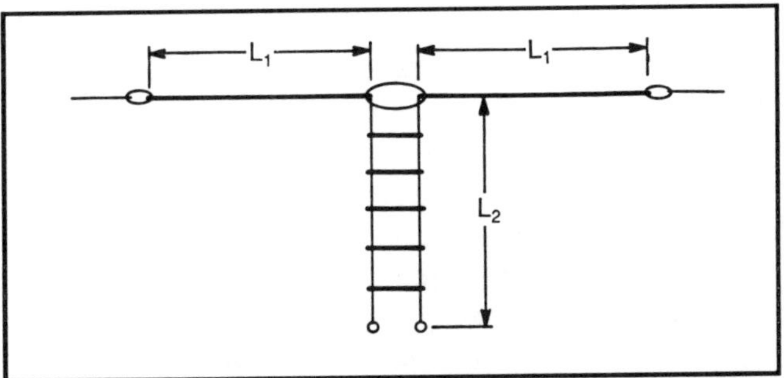

Figure 3-7. The symbols used for the dimensions of a center-fed dipole with open-wire feeders (or twin-lead with appropriate consideration of velocity factors).

Sec 3.5.1 The "Worst Case" Design

Apparently, an 80-meter dipole with a quarter-wave open-wire (or twin-lead) feedline is a logical design for a multiband dipole. In **Figure 3-7** this would mean that $L_1 = L_2 = 59$ to 67 feet, depending upon the favorite operating frequency. If one were to use 450-ohm twin-lead with "open windows," L_2 would be diminished by 10 percent; with 300-ohm TV twin-lead, it would be diminished by about 20 percent. This would make a good antenna system on 160 meters. Because $L_1 + L_2$ is close to a quarter-wave, the current at the input to the feedline is at its highest value and the impedance at its lowest—a very favorable condition for the balun. In fact, a 1:1 balun at the feedpoint would probably do a good job.

However, what does the input to the feedline look like on 80 meters? If 450-ohm feedline is used, the 4:1 balun sees a quarter-wave 450-ohm feedline terminated in approximately 50 ohms. Transmission line theory tells us that the balun would see 4050 ohms—an impossible condition for most baluns. The situation becomes worse on 40 meters, where we may have a center-fed, full-wave dipole with a half-wave transmission line (a 1:1 matching transformer). In this case, the balun could see an impedance approaching 10,000 ohms—a more than impossible condition! Although high impedances would also be seen on 10, 15 and 20 meters, the conditions are not quite as severe because the balun's choking reactances are usually greater and the impedances lower. In essence, the "worst case" design for a 4:1 balun is the "best case" design (except for 160 meters), if one still uses inductive coupling to a parallel-tuned circuit.[19]

Sec 3.5.2 A Smaller Design— The G5RV

Varney,[24] G5RV, designed a multiband center-fed antenna system capable of operation on all HF bands from 3.5 to 30 MHz. In contrast to multiband antennas designed as half-wave dipoles on 80 meters (the "worst case" design), the full-size G5RV antenna was designed as a three-half-wave antenna on 14,150 MHz with a 1:1 transmission line matching transformer. It was possible to accomplish this using the dimensions of 51 feet for L_1 and 34 feet for L_2. For 450-ohm twin-lead with "windows," L_2 would be 31 feet; for 300-ohm TV ribbon, L_2 would be 28 feet. Consequently, the input impedance at the base of the matching transmission line was about 100 ohms on 14,150 MHz, and a manageable impedance for 50 or 80-ohm coaxial cable.

However, with a total length for $L_1 + L_2$ of 85 feet, the impedances at the input to the transmission line on 40, 80, and 160 meters are also manageable. Even though they have a reactive component on these bands, they aren't so high that a well-designed 4:1 balun in an antenna tuner can't handle them easily. Varney also showed that the highest impedances occurred on the 18, 21, and 28 MHz bands. However, this doesn't present a problem with the design shown in **Photo 3-D** because it uses an efficient core material, and it also has the highest reactances of its windings at these frequencies.

Varney[24] also wrote, at considerable length, on the unsuitability of a balun used to connect the base of the 34-foot open feeders to a coaxial cable feedline. He states that if a balun is connected to a reactive load with a VSWR of more than 2:1, its internal losses increase. Varney also mentioned heating of the wires and saturation of the core. Evidently Varney was not familiar with McCoy's design, which uses a powdered-iron core (with a permeability of 10) that can withstand VSWRs considerably greater than 2:1—without showing any temperature rise. Furthermore, the wire doesn't heat up; however, the core does via dielectric heating. Additionally, with sufficient choking reactance, baluns can handle (equally) the resistive and reactive components of an impedance.

Finally, after observing the voltage and current distributions on all of the bands, it appears that a 2:1 (100:50-ohm) balun might be an interesting one to try on the G5RV antenna. It could be that many of the bands would not require the added matching of an antenna tuner. If some of the bands require an antenna tuner in order to be used, then I would suggest using a "hardened" balun. That is what I call McCoy's approach, which uses efficient and hardy powdered-iron cores. A 2:1 balun, comprised of a 1:2 unun in series with a 1:1 Guanella (current) balun, could be easily designed and built (see **Chapter 4**).

Sec 3.5.3 A Larger Design

Even though the G5RV antenna can be made to operate on 160 meters with a suitable antenna tuner, an antenna system larger than the "worst case" design can provide better operation on the 40, 80, and 160 meter bands. As you might expect, an antenna system about twice as large as the G5RV offers these advantages. Suggested dimensions are L_1 = 80 feet and L_2 = 100 feet. If the feedline is 450-ohm twin-line with "windows," then L_2 = 90 feet. If it is 300-ohm TV ribbon, then L_2 = 82 feet. As with the G5RV, it's the total length of $L_1 + L_2$ that presents favorable or unfavorable impedances to the 4:1 balun in the antenna tuner. Therefore, L_1 and L_2 could both be 90 feet, as well. Only very small differences in performance would be noticed between these two systems, particularly on the lower-frequency bands. Obviously, other combinations totaling 180 feet are also possible. In the G5RV case, it's 85 feet.

Sec 3.6 Summary

After reading this chapter, one might think I have set this technology back a few years by advocating voltage baluns and powdered-iron cores. I have even questioned the professional literature. However, my conclusions were based upon three experimental results. These were: 1) measurements with my resistive bridge on input impedances of 4:1 Ruthroff (voltage) baluns with loads floating and center-tapped-to-ground, 2) VSWR measurements on folded dipoles with various 4:1 baluns (large and small and therefore with many different lengths of transmission lines), and 3) McCoy's success with his 4:1 balun. Also, as this chapter points out, it helps to have the dimensions of a multiband, center-fed dipole, and feeders favor the operation of a 4:1 balun in antenna tuners. And, *yes*, there is a "worst case" antenna design!

As in many investigations, supplying answers to some questions can lead to others that appear to be important. Specifically, for powdered-irons, how would permeabilities in the 20 to 35 range perform? Simple impedance measurements showing lower values on input impedances, indicate that there is more loss than with a permeability of 10; but accurate insertion loss measurements are needed in order to tell the complete story—the trade-off in efficiency for low-frequency response.

Finally, low permeability ferrite-like No. 67 material with a permeability of 40 looks interesting for use in baluns for antenna tuners. Accurate insertion loss measurements[2] have also shown the very same high efficiency that was exhibited by powdered-iron having a permeability of 10. With a sufficient number of turns on an appropriate size core, a balun made of this material could be practical. Even though the amateur radio literature still refers to the problem of core saturation, there has been only one recorded case. This was on 2 MHz with a rod-type 1:1 balun where insufficient choking reactance exists.[13,14]

Chapter 4

1.5:1 and 2:1 Baluns

Sec 4.1 Introduction

There are many applications for broadband baluns with impedance transformation ratios close t 1.5:1 and 2:1. Two applications involve matching 50-ohm cable to balanced loads of 75 or 100 ohms, which are the input impedances of a half-wave dipole at heights of 0.22 or 0.34 wavelengths above ground. Another, is the matching of 50-ohm cable to the 100-ohm input impedance of a quad antenna. An interesting, and somewhat unexpected application, is the matching of 50-ohm cable directly to the input impedance of the driven element of a Yagi beam antenna of 33 or 25 ohms. This would eliminate the common hairpin matching network presently used to raise their input impedances to 50 ohms.

There are many versions of these two baluns. They include: 1) high- and low-power designs, 2) designs matching 50-ohm cable to higher or lower impedances, 3) series- or parallel-type designs, 4) single- or dual-core designs, 5) dual-ratio designs, and 6) HF and VHF designs. The series-type baluns use an unun (unbalanced-to-unbalanced transformer) in series with a Guanella (current) balun. More details on these ununs are provided in later chapters. In this chapter, you'll read about many high-power designs capable of handling the full legal limit of amateur radio power. They are optimized for sufficient margins in choking reactance at their low frequency ends, and in efficiency throughout their passbands. Two of the 2:1 baluns are specifically designed for 2-meter operation.

Sec 4.2 1.5:1 Baluns

In this section, I'll present two series-type 1.5:1 baluns (actually 1.56:1, which should be close enough). They both use 1.56:1 ununs in series with Guanella 1:1 baluns. **Figures 4-1A** and **B** show their schematic diagrams. **Figure 4-1B** has an extra input (to a tap), which provides another ratio of 1.33:1.

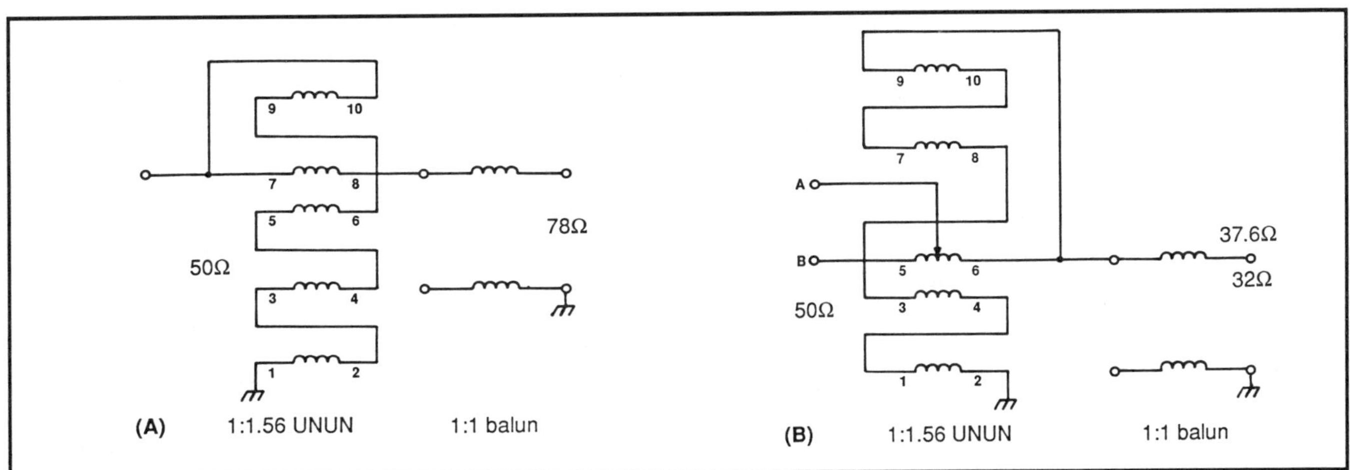

Figure 4-1. Schematic diagrams of two 1.56:1 baluns: (A) step-up, 50:78 ohms; (B) step-down, 50:37.6 ohms—connection A, 50:32 ohms—connection B.

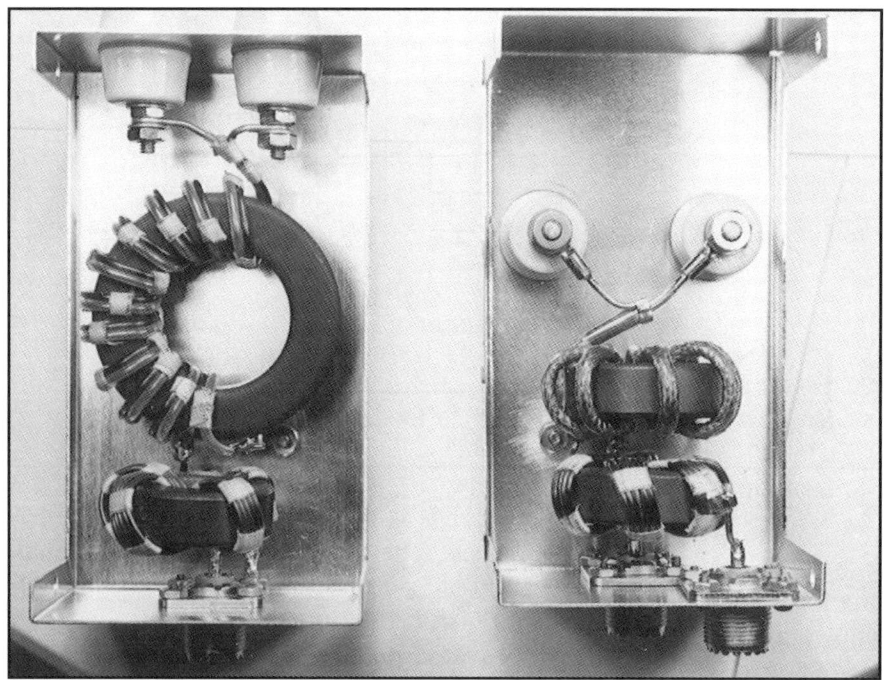

Photo 4-A. Baluns using the schematic diagrams of Figure 4-1. Balun on the left matches 50-ohm cable to a balanced load of 78 ohms. Balun on the right matches 50-ohm cable to balanced loads of 37.6 or 32 ohms.

The left-hand side of **Photo 4-A** shows a design using **Figure 4-1A** mounted in a CU 3006 minibox 5.25 inches long by 3 inches wide by 2.25 inches high (Radio Shack carries a similar enclosure). The 1:1.56 unun has 4 quintufilar turns on a 1.5-inch OD ferrite toroid with a permeability of 250. Winding 7-8 is No. 14 H Thermaleze wire and the other four are No. 16 H Thermaleze wire.

The 1:1 Guanella balun has 11 bifilar turns of No. 14 H Thermaleze wire on a 2.4-inch OD ferrite toroid with a permeability of 250. One wire is covered with Teflon^{TM} tubing, resulting in a characteristic impedance very close to 78 ohms (the optimum value).

When matching 50-ohm cable to a balanced load of 78 ohms, the impedance transformation ratio is literally flat (within a percent or two) from 1.5 MHz to 40 MHz! You might be interested to know that (separately) the 1:1 (75:75 ohm) balun would make an excellent isolation transformer for 75-ohm hardline, and the 1.56:1 (78:50 ohm) unun an excellent match between 75-ohm hardline and 50-ohm cable.

The right-hand side of **Photo 4-A** shows a design using **Figure 4-1B** mounted in a similar enclosure. The 1.56:1 unun has 5 quintufilar turns on a 1.5-inch OD ferrite toroid with a permeability of 250. Winding 5-6 is No. 14 H Thermaleze wire and is tapped at one turn from terminal 5. The other four wires are No. 16 H Thermaleze.

The 1:1 Guanella balun has 7 turns of homemade coaxial cable on a 1.5-inch OD ferrite toroid with a permeability of 250. The inner conductor is No. 14 H Thermaleze wire and is covered with Teflon tubing. The outer braid, which is from a small coaxial cable or from 1/8-inch tubular braid, is also tightly wrapped with Scotch No. 92 tape to preserve the low characteristic impedance.

In matching 50-ohm cable to a balanced load of 37.6 ohms (connection A), or to a balanced load of 32 ohms (connection B), the response is essentially flat (within a percent or two) from 1.5 to 30 MHz.

Sec 4.3 2:1 Baluns

The 2:1 balun lends itself to more choices in design than the 1.56:1 balun. This is especially true because the parallel-type design, which provides a 2.25:1 balun with the widest possible bandwidth, can easily be employed. The 1.56:1 balun is at a disadvantage here. This section presents many baluns using both series and parallel-type designs.

Sec 4.3.1 Series-type Baluns

Figure 4-2 shows circuit diagrams for two versions of the series-type balun. **Photo 4-B** shows a design using **Figure 4-2A** mounted in a CU 3005-A minibox 5 inches long by 4 inches wide by 3 inches high. The 1:2 unun has 7 trifilar turns on a 1.5-inch OD ferrite toroid with a permeability of 250. The output tap is located 6 turns from terminal 5. Winding 5-6 is No.

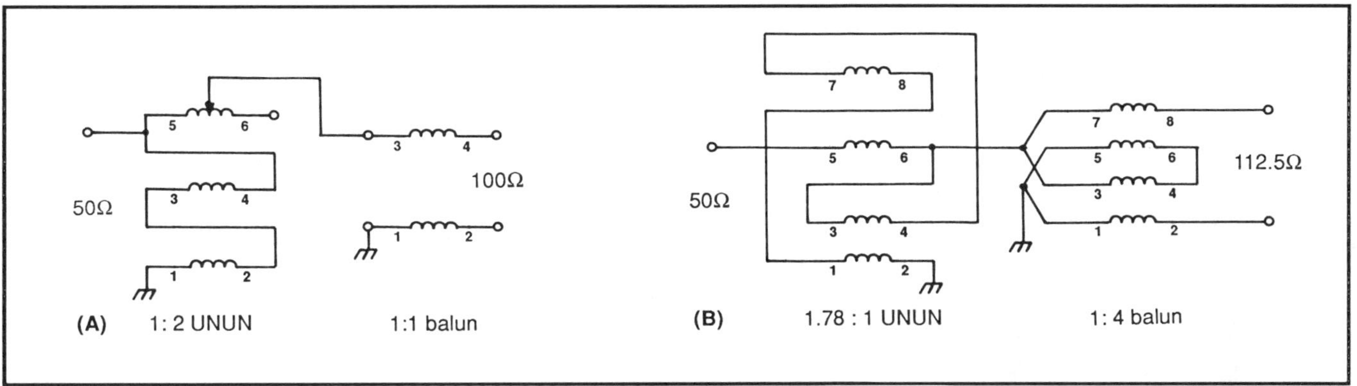

Figure 4-2. Schematic diagrams of two versions of the series-type balun: (A) 1:2 (50:100 ohms) balun; (B) 1:2.25 (50:112.5 ohms) balun.

14 H Thermaleze wire and the other two are No. 16 H Thermaleze wire.

The 1:1 Guanella balun has 14 bifilar turns of No. 14 H Thermaleze wire on a 2.4-inch OD ferrite toroid with a permeability of 250. Both wires are covered with Teflon tubing, which results in a characteristic impedance of 100 ohms (the optimum value). A crossover, placing 7 turns on one side of the toroid and 7 turns on the other, is used so the output and input are on opposite sides of the toroid. **Figure 4-3** is a drawing of the crossover. Although this technique has no electrical advantage at HF, the mechanical advantage is obvious.

When matching 50-ohm cable to a balanced load of 100 ohms, the response is literally flat (within 2 to 3 percent) from 1.5 to 30 MHz. By connecting the output of the unun to terminal 6 instead of to the tap, the balun would match 50-ohm cable to a balanced load of 112.5 ohms with about the same response.

Photo 4-B. A series-type balun using the schematic diagram of Figure 4-2A designed to match 50-ohm cable to a balanced load of 100 ohms.

Photo 4-C shows two slightly different versions of series-type 2.25:1 baluns using the circuit of **Figure 4-2B**. Both have the same 1.78:1 step-down unun, which has 5 quadrifilar turns on a 1.5-inch OD ferrite toroid with a permeability of 250. Winding 5-6 is No. 14 H Thermaleze wire, and the other three are No. 16 H Thermaleze wire. Each version also has 8 bifilar turns of No. 14 H Thermaleze wire on both of the 1.5-inch OD ferrite toroids, with a permeability of 250.

The differences are: 1) the balun on the left in **Photo 4-C** has one layer of Scotch No. 92 tape on one of the wires in each bifilar winding and a crossover after the fourth turn, and 2) the balun on the right has two layers of Scotch No. 92 tape on one of the wires on one toroid and no extra insulation on the wires of the other toroid. Therefore, one of the windings in the 1:4 Guanella balun has a characteristic impedance a little less than 50 ohms and the other a little greater than 50 ohms, resulting in a canceling effect. Furthermore, the crossover isn't used in this design. The balun on the left is mounted in a CU 3006 minibox 5.25 inches long by 3 inches wide by 2.25 inches high. The balun on the right is mounted in a CU 3015-A minibox 4 inches long by 2 inches wide by 2.75 inches high.

The performance of these two baluns is essentially the same. When matching 50-ohm cable to balanced loads of 112.5 ohms, the responses are essentially flat (within 2 to 3 percent) from 1.5 to 30 MHz.

From preliminary measurements on series-type 2:1 baluns, the balun in **Photo 4-B** is the one I'd recommend for matching 50-ohm cable to balanced loads of 100 ohms, while the baluns in **Photo 4-C** would be best for matching to balanced loads of 112.5 ohms.

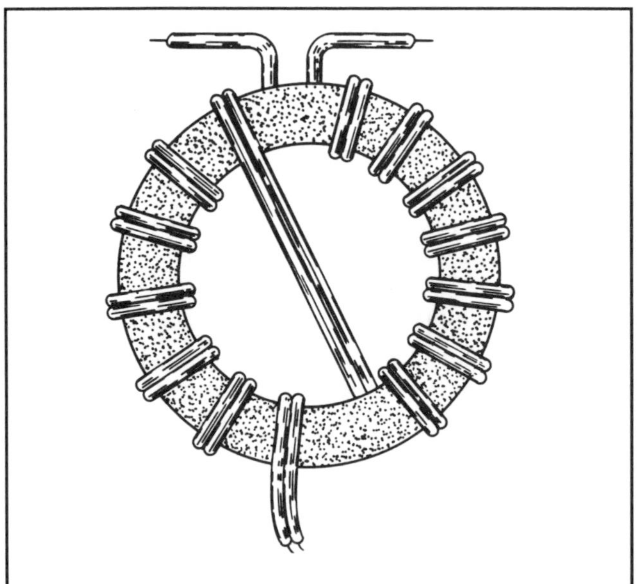

Figure 4-3. Construction of a 1:1 Guanella balun with a crossover placing the input and output terminals on opposite sides of the toroid.

Photo 4-C. Two series-type baluns using the schematic diagram of Figure 4-2B designed to match 50-ohm cable to balanced loads of 112.5 ohms.

Also, by replacing the 1.78:1 unun in **Figure 4-2B** with a 2.25:1 unun, and not adding any extra insulation to the windings of the 1:4 balun, it's possible to obtain an excellent balun matching 50-ohm cable to a balanced load of 89 ohms.

Figure 4-4 is the schematic diagram of a series-type balun designed to match 50-ohm cable to balanced loads of 25 or 22.22 ohms. **Photo 4-D** shows two versions of this dual-ratio balun. The balun on the left, for its unun, has 6 trifilar turns of No. 14 H Thermaleze wire on a 1.5-inch OD ferrite toroid with a permeability of 250. Winding 3-4 is tapped one turn from terminal 3, yielding the 2:1 ratio. The 1:1 Guanella balun has 6 turns of homemade coaxial cable on a similar toroid. The inner conductor is No. 12 H Thermaleze wire and is covered with Teflon tubing. The outer braid, from a small coax or from 1/8-inch tubular braid, is tightly wrapped with Scotch No. 92 tape to preserve its low characteristic impedance.

In matching 50-ohm cable to balanced loads of 25 ohms (connection A) or to 22.22 ohms (connection B), the response is essentially flat (within a percent or two) from 1.5 to 30 MHz.

The balun on the right in **Photo 4-D** has similar windings on a single 2.4-inch OD ferrite toroid with a permeability of 250. Its performance is quite comparable to the balun on the left.

Sec 4.3.2 Parallel-type Baluns

As you saw in the previous section, the series-type baluns presented here are combinations of ununs with ratios of 1.33:1, 1.78:1, 2:1, and 2.25:1 in series with Guanella 1:1 or 4:1 baluns. The ununs, which are really an extension of Ruthroff's[9] bootstrap technique for obtaining a 4:1 unun, sum direct voltages with delayed voltages that traverse a single transmission line. Therefore, the unun eventually limits the high-frequency response of the series-type balun.

On the other hand, the parallel-type balun is an extension of Guanella's technique of summing voltages of equal delays. Instead of simply connecting transmission lines in parallel-series, the parallel-type balun connects Guanella baluns in parallel-series. As I noted in **Reference 25**, two 4:1 Guanella baluns can be connected in parallel-series, yielding very broadband ratios of 6.25:1. This section shows how a 1:1 Guanella balun can be connected with a 4:1 Guanella balun in parallel-series, yielding a very broadband ratio of 2.25:1.

Figure 4-5 is the schematic diagram of the high-frequency model of a 2.25:1 unun which is used for analysis purposes. Because the current through the load is $3/2I_1$, the transformation ratio is $(3/2)^2$, or 2.25:1. Therefore, if the impedance seen on the left side is 50 ohms, a matched impedance on the right side is 22.22 ohms. Because two thirds of the 50 ohms appears across the input of the Guanella 1:1 balun, its optimum characteristic impedance is 33.33 ohms. Similarly, this is also the value of the optimum characteristic impedance for the windings of the 4:1 balun. Because the 1:1 balun wants to see 33.33 ohms on its output (a matched condition) and the 4:1 balun wants to see 66.66 ohms, placing these two values in parallel results in the confirming value of 22.22 ohms.

If the 50-ohm generator is placed on the right side in **Figure 4-5**, the circuit becomes a step-up unun matching 50 ohms to 112.5 ohms (on the left). If the ground is removed on the left side, the transformer becomes a balun. A similar analysis as above, shows that the optimum characteristic impedance of the three bifilar windings now becomes 75 ohms.

Photo 4-E shows a parallel-type 2.25:1 balun designed to match 50-ohm cable to a balanced load of 112.5 ohms. It has 9 bifilar turns of No. 14 H Thermaleze wire on each of the three toroids that have a 1.5-inch OD and a permeability of 250. Also, one of the wires on each toroid is covered with Teflon tubing, resulting in a characteristic impedance of 75 ohms (the optimum value). When operating as a balun, the response is essentially flat from 7 MHz to over 45 MHz. As an unun, the flat response is broadened from 1.5 MHz to over 45 MHz.

Because the coiled wire, parallel-type balun didn't provide any real advantage over the series-type balun (in fact, the low-frequency response was poorer), I investigated the beaded transmission line version for possible use in the VHF band. **Figure 4-6** shows a schematic diagram of one using coaxial cable. Obviously, twin lead could be substituted for the coaxial cable. **Photo 4-F** shows both versions.

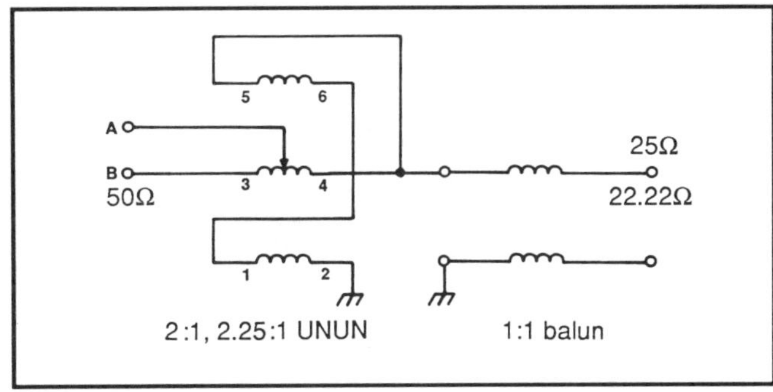

Figure 4-4. Schematic diagram of a dual-ratio series-type balun. Connection A matches 50-ohm cable to a balanced load of 25 ohms. Connection B matches 50-ohm cable to a balanced load of 22.22 ohms.

The top design in **Photo 4-F** has 4 inches of 3/8-inch OD ferrite beads with a permeability of 125 on each of the three 75-ohm transmission lines. It is designed to match 50-ohm cable on the right to a balanced load of 112.5 ohms on the left (with the ground on terminal 1 removed). The transmission lines consist of two No. 14 H Thermaleze wires separated by the Teflon tubing covering one of them. When matching 50-ohm cable to a balanced load of 112.5 ohms, the response is essentially flat from 30 MHz to over 100 MHz (the limit of my bridge).

The bottom design in **Photo 4-F** also has 4 inches of 3/8-inch OD ferrite beads with a permeability of 250. However, they are now threaded by homemade

Photo 4-D. Two series-type baluns using the schematic diagram of Figure 4-4 designed to match 50-ohm cable to balanced loads of 25 ohms or 22.22 ohms.

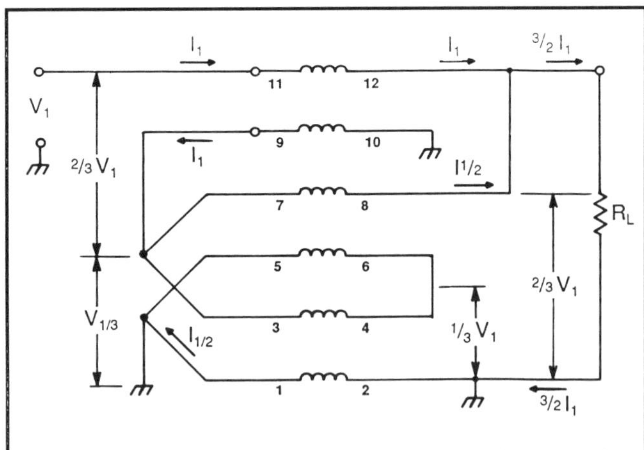

Figure 4-5. High-frequency model of the parallel-type 2.25:1 transformer. Connections shown are for unun operation.

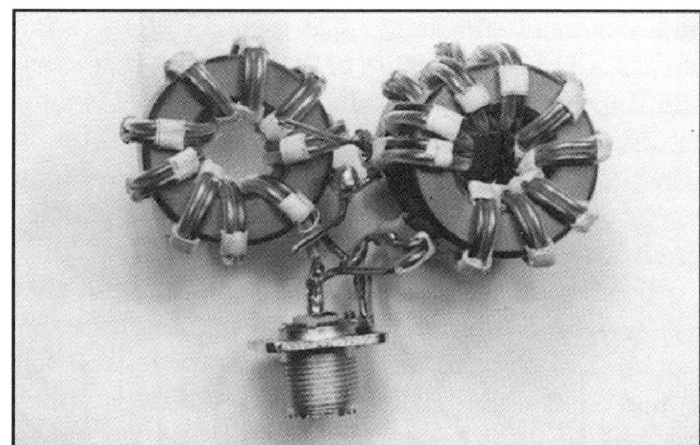

Photo 4-E. Parallel-type 2.25:1 balun matching 50-ohm cable to a balanced load of 112.5 ohms.

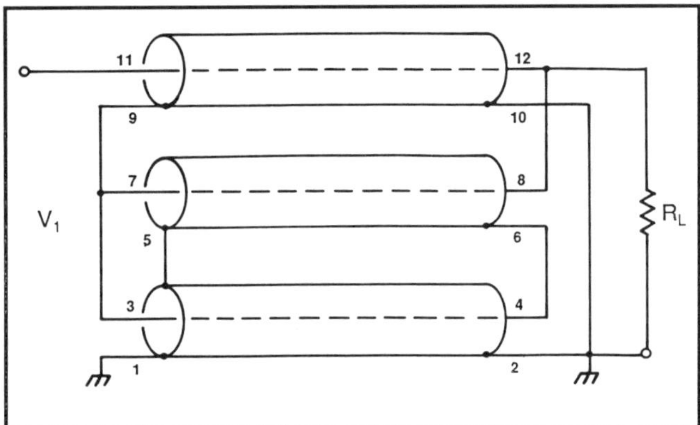

Figure 4-6. The coaxial cable version of the parallel-type 2.25:1 transformer of Figure 4-5.

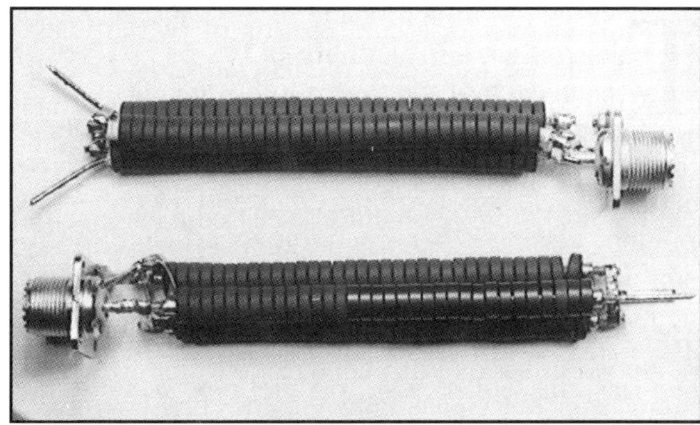

Photo 4-F. Beaded transmission line versions of the 2.25:1 parallel-type balun for operation in the VHF band. The top balun matches 50-ohm cable to a balanced load of 112.5 ohms. The bottom balun matches it to a balanced load of 22.22 ohms.

coaxial cable with a characteristic impedance of 33.33 ohms. It is designed to match 50-ohm cable on the left to a balanced load of 22.22 ohms (with the ground on terminal 2 removed). The inner conductor of the coax is No. 14 H Thermaleze wire and is covered with Teflon tubing. The braid is from small coaxial cable or from 1/8-inch tubular braid. The braid is also tightly wrapped with Scotch No. 92 tape in order to preserve the low characteristic impedance. When matching 50-ohm cable to a balanced load of 22.22 ohms, the response is essentially flat from 14 MHz to over 100 MHz (again, the limit of my bridge).

Sec 4.4 Closing Comments

In closing, I'd like to make a couple of comments regarding parallel-type baluns.

First, if you are interested in a 1.78:1 ratio, replace the 1:4 balun in **Figures 4-5** and **4-6** with a 1:9 Guanella balun (three transmission lines connected in parallel-series). This would yield an output current of $4/3I_1$ and a ratio of $(4/3)^2$, or 1.78:1. If you replace the 1:4 balun with a 1:16 Guanella balun (four transmission lines connected in parallel-series), the output current will be $5/4I_1$, with a ratio of $(5/4)^2$, or 1.56:1.

Second, because the parallel-type balun is really an extension of Guanella's technique of summing voltages of equal delays,[26,27] the high-frequency response is mainly limited by the parasitics in the interconnections. Therefore, beaded transmission lines offer the best opportunity for successful operation on the VHF band. It is also recommended that the ferrite beads have permeabilities of 300 or less[2] in order to achieve the very high efficiencies of which these transformers are capable.

Chapter 5

6:1, 9:1 and 12:1 Baluns

Sec 5.1 Introduction

This chapter is a combination of two articles I wrote for *Communications Quarterly*. One, on the 6:1 and 9:1 baluns, appeared in the Winter 1993 issue and the other, on the 12:1 balun, ran in the Summer 1993 issue. I combined these two articles here because the three baluns discussed have much in common. The commonalities include: 1) They are the most difficult to construct and are therefore the most expensive; 2) two of them, the 6:1 and 12:1 baluns, use ununs in series with Guanella baluns; 3) two of them use the 9:1 Guanella balun—the 12:1 balun with a series 1.33:1 unun; 4) they are generally associated with two-wire transmission lines with characteristic impedances of 300, 450, and 600 ohms, respectively; 5) when matching 50-ohm cable to higher impedances, they have more loss than the baluns in the preceding chapters; and 6) the optimum application of these baluns requires a more critical understanding of the trade-offs in low-frequency response for efficiency.

Like the 2:1 balun in **Chapter 4**, the 6:1 (actually 6.25:1) balun also comes in two forms: the series-type, which offers better low-frequency response in the HF band, and the parallel-type, which has a vastly greater high-frequency capability. The parallel-type 6:1 balun together with the 9:1 Guanella balun (which is also a parallel-type) offers the potential for designs capable of efficient and broadband operation on the VHF and UHF bands.

Sec 5.2 6:1 and 9:1 Baluns

Many radio amateurs associate the use of the 6:1 and 9:1 baluns with 300-ohm twin lead feeding folded dipoles, or 450-ohm "ladder" line feeding single or multi-band antenna systems. However, what is neglected (in some cases) is the effect of the height of these antennas above earth and the length of the transmission lines feeding them.

Broadband 6:1 and 9:1 baluns are considerably more difficult to construct than the more common 1:1 and 4:1 baluns. This is especially true when matching 50-ohm cable to impedances of 300 and 450 ohms. Furthermore, there are some important trade-offs in low-frequency response for efficiency.

From what I could gather "on-the-air" or talking to radio clubs, I have determined what I believe are probably two of the most common misconceptions regarding the use of these baluns:

1. 6:1 Baluns. In free-space, the folded dipole with 300-ohm twin-lead has a resonant impedance close to 300 ohms. The dipole also has this value at a height of about 0.225 wavelength above ground. However, it's only 200 ohms at a height of about 0.17 wavelength and 400 ohms (the maximum) and at 0.35 wavelength. In many cases, the 4:1 balun would actually do a better job of matching.

2. 9:1 Baluns. Some are unaware of the relationship between the impedance at the input of a transmission line, the characteristic impedance of the line, and the impedance at the end of the line. Just because a transmission line has a characteristic impedance of 450 ohms doesn't necessarily mean that a 9:1 balun will perform a satisfactory match to 50-ohm cable. Far from it. For example, if the line is terminated by a half-wave dipole with an impedance of 50 ohms, the 9:1 balun would see 50 ohms when the line is a half-wave long and 4050 ohms when it's a quarter-wave long! Broadband baluns cannot be designed to handle impedances as high as 4050 ohms. It's very likely that a well-designed 50:450-ohm balun would experience (particularly on 80 and 160 meters) harmful flux in

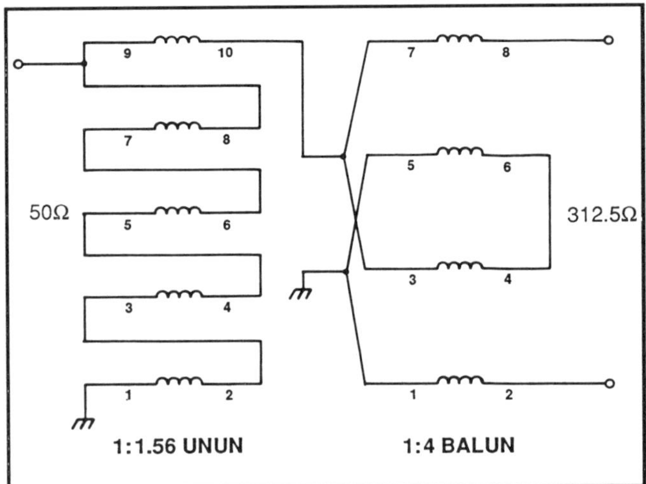

Figure 5-1. Schematic diagram of the series-type balun with a 1:6.25 ratio designed to match 50 to 312.5 ohms.

the core *and* excessive heating because of the large voltage drop along the length of its transmission lines. This problem of presenting very high (and harmful) impedances to baluns is quite prevalent with multi-band antenna systems.

Clearly, there are many applications for 6:1 and 9:1 baluns. They not only include matching 50-ohm cable to balanced loads of 300 and 450 ohms, but also to balanced loads of 8 and 5.6 ohms, as well. Furthermore, many of the designs in this chapter will perform almost as well in unun (unbalanced-to-unbalanced) applications. The trade-off (which is usually very small) is in low-frequency response. Additionally, these baluns could be used to exploit the low-loss properties of 300 and 450-ohm twin-lead where very long transmission lines are used. This is especially important at 14 MHz and above.

In the pages that follow, I'll present a variety of baluns matching 50-ohm cable to 300 ohms (actually to 312.5 ohms, a 6.25:1 ratio) and to 450 ohms, as well as 50-ohm cable to 8 and 5.6 ohms. Also included are two different versions of 6.25:1 baluns. One is a series-type using a 1.56:1 unun in series with a 4:1 Guanella balun and the other a series-parallel arrangement using two 4:1 Guanella baluns. Because the series-parallel balun adds voltages of equal delays, you'll find its high-frequency capability is much greater.

The 9:1 balun is a conventional Guanella balun with three transmission lines connected in series at the high-impedance side and in parallel at the low-impedance side. Therefore, it also sums voltages of equal delays. Some of the comparisons and analyses of these 6.25:1 and 9:1 baluns were probably published for the first time in my Winter 1993 *Communications Quarterly* article.

Sec 5.2.1 6.25:1 Series-Type Baluns

Figure 5-1 shows the schematic diagram of a series-type balun designed to match 50-ohm cable to a balanced load of 312.5 ohms. It consists of a 1:1.56 unun in series with a 1:4 Guanella balun. The overall ratio of 1:6.25 should satisfy most of the 1:6 requirements. **Photo 5-A** shows three examples. All three baluns use the same step-up unun that has four quintufilar turns on a 1.5-inch OD ferrite toroid with a permeability of 250. Winding 9-10 is No. 14 H Thermaleze wire and the other four are No. 16 H Thermaleze wire. Because this unun sums only one delayed voltage with four equal direct voltages, it has an excellent high-frequency response.[2]

The balun on the left in **Photo 5-A** has eight bifilar turns of No. 18 hook-up wire on each transmission line of its 1:4 balun. The wires are further spaced with No. 18 Teflon™ tubing providing a characteristic

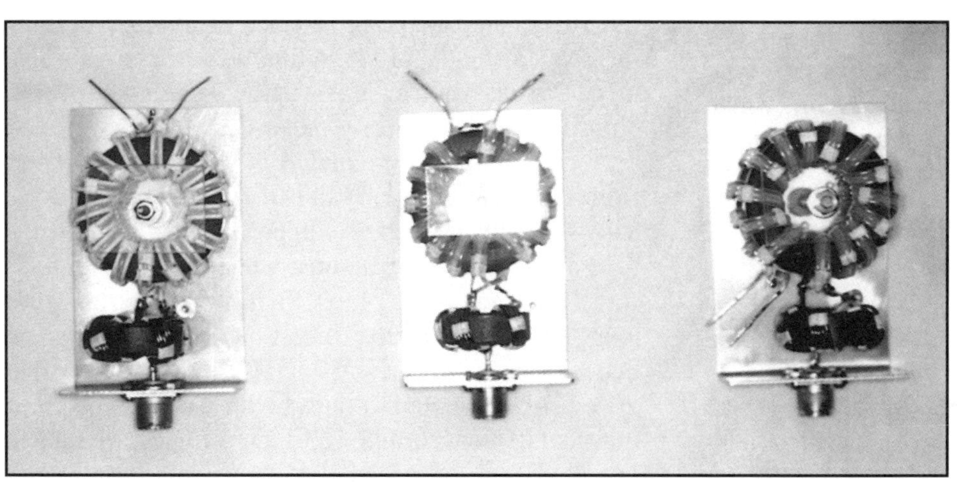

Photo 5-A. Three examples of series-type 1:6.25 baluns. The balun on the right, with a double-core 1:4 balun, has both a balanced voltage and current output. The other two only have balanced-current outputs.

impedance close to 150 ohms (the optimum value). The ferrite toroid has a 2.4-inch OD and a permeability of 250. When matching 50-ohm cable to a floating load of 312.5 ohms, the response is essentially flat from 1.7 to 30 MHz. Under matched conditions, 500 watts of continuous power and 1 kW of peak power is a conservative power rating. Because the 1:4 balun in this series-type 1:6.25 balun uses only one core instead of two, this transformer should never be used when the load is grounded at its center. Also, it is not recommended for balanced antennas. This series-type balun presents balanced currents, but does *not* present balanced voltages.

The balun in the center of **Photo 5-A** has seven bifilar turns of No. 16 SF Formvar™ wire on each transmission line on its 1:4 balun. The wires are covered with Telfon sleeving and further separated by No. 16 Teflon tubing. Like the balun on the left, the characteristic impedance is also close to the objective of 150 ohms. The toroid also has a 2.4-inch OD and a permeability of 250. When matching 50-ohm cable to a floating load of 312.5 ohms, the response is essentially flat from 3.5 to 30 MHz. Over this frequency range, this balun can easily handle the full legal limit of amateur radio power. Because this balun also presents balanced currents and not balanced voltages, it should not be used when the loads are balanced to ground or grounded at their centers.

The balun on the right in **Photo 5-A** has fourteen bifilar turns of No. 16 SF Formvar wire on each of the two toroids of the 1:4 balun. The wires are also covered with Teflon sleeving and further separated by No. 16 Teflon tubing. For ease of connection, one core is wound clockwise and the other counterclockwise. The two cores are spaced 1/4 inch apart with acrylic sections. When matching 50-ohm cable to a 312.5 load that is either floating, balanced to ground, grounded at its center, or grounded at its bottom (a broadband unun), the response is essentially flat from 1.7 to 30 MHz. Under this matched condition, it can easily handle the full legal limit of amateur radio power. Furthermore, this is a true balun because it presents equal currents and equal voltages. If one were to measure the voltages-to-ground, when matching into a balanced load, one would find them to be equal and opposite. The other two 1:6.25 baluns using single-core 1:4 baluns, would have equal currents but not equal voltages (see **Chapter 2**). Because they are easier to construct, it would be interesting to compare them with a true balun.

Figure 5-2 is the schematic diagram of a series-type balun designed to match 50-ohm cable to a balanced load of 8 ohms (perhaps a short-boom Yagi). It consists of a 1.56:1 step-down unun in series with a Guanella 4:1 step-down balun. The overall ratio is 6.25:1. **Photo 5-B** shows two examples. Both baluns use the same step-down unun, which has four quintufilar turns on a 1.5-inch OD ferrite toroid with a permeability of 250. Winding 5-6 is No. 14 H Thermaleze wire and the other four are No. 16 H Thermaleze wire. The interleaving of the wires is such that the

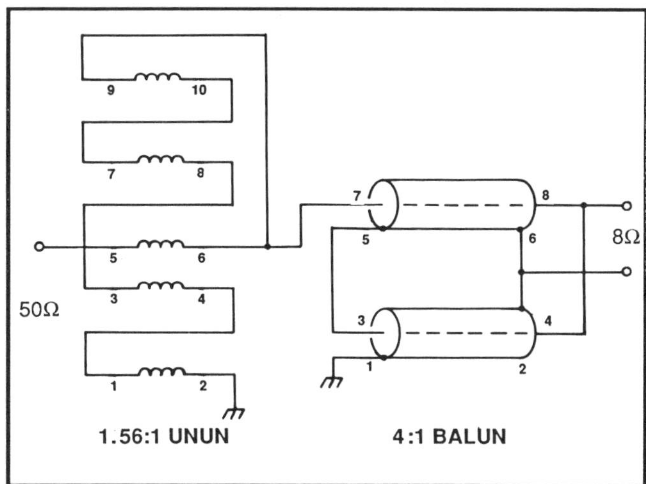

Figure 5-2. Schematic diagram of the series-type balun with a 6.25:1 ratio designed to match 50 to 8 ohms.

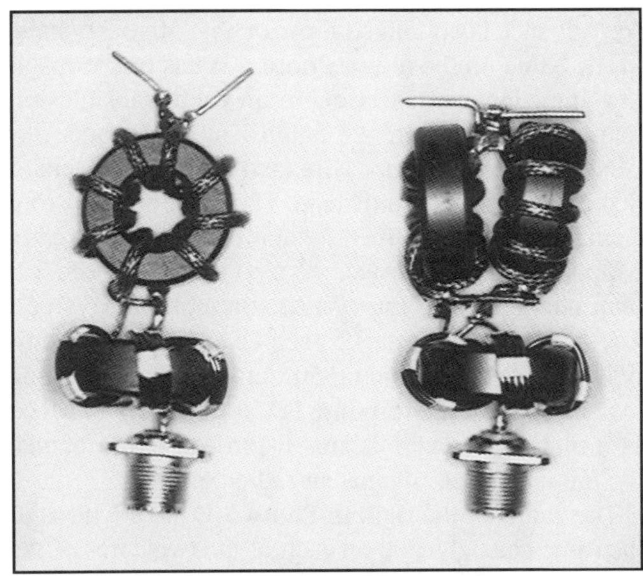

Photo 5-B. Two examples of the series-type 6.25:1 balun optimized at the 50:8-ohm level. The balun on the left is designed to match into a floating 8-ohm load. The balun on the right is designed to match into an 8-ohm floating, center-tapped-to-ground or grounded load (unun).

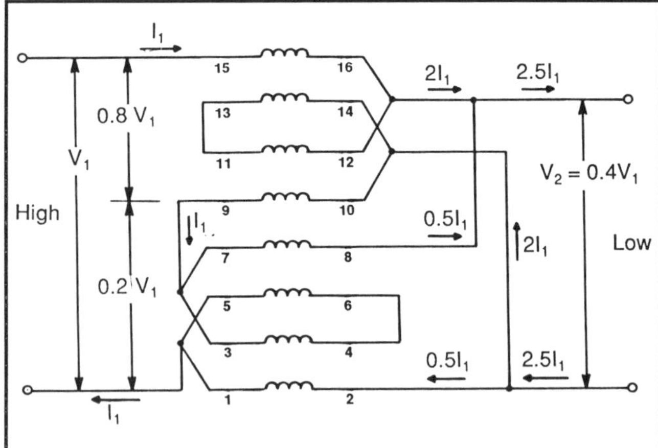

Figure 5-3. Schematic diagram of the parallel-type balun (and unun) with a 6.25:1 ratio. The currents and voltages are shown for analysis purposes (see text).

performance is optimized for matching 50 to 32 ohms.

The balun on the left in **Photo 5-B** has four turns of low-impedance coaxial cable on each transmission line on the single-core 4:1 balun. The inner conductor is No. 14 H Thermaleze wire, and it has two layers of Scotch No. 92 polyimide tape. The outer braid is from a small coax (or 1/8-inch tubular braid) and is tightly wrapped with Scotch No. 92 tape to achieve the 17-ohm characteristic impedance (the optimum value). The ferrite toroid has a 1.5-inch OD and a permeability of 250. When matching 50-ohm cable to a floating load of 8 ohms, the response is flat from 1 to 40 MHz. In a matched condition, this balun can easily handle the full legal limit of amateur radio power.

The balun on the right in **Photo 5-B** has six turns of the same coaxial cable on each of the two cores of the Guanella step-down balun. The cores also have a 1.5-inch OD and a permeability of 250. The performance of this balun is practically the same as the above with the single-core 4:1 balun. The important differences are that this 6.25:1 balun also performs equally well whether the load is center-tapped-to-ground, balanced-to-ground, or grounded at the bottom (a broadband unun). This is the one recommended for feeding a short-spaced Yagi beam antenna.

Sec 5.2.2 6.25:1 Parallel-Type Baluns

The 6.25:1 series-type baluns described in the preceding section consisted of a 1.56:1 unun, which is an extension of Ruthroff's "bootstrap" approach for ununs,[9] in series with a Guanella 4:1 balun.[2] The upper-frequency limit for this combination is really set by the unun, which sums a delayed voltage with four direct voltages. The parallel-type 6.25:1 baluns described in this section are really extensions of Guanella's approach, which sums voltages of equal delays. Therefore, the upper-frequency limit is mainly dependent upon the parasitics in the interconnections.

The 6.25:1 parallel-type balun uses two 4:1 Guanella baluns connected in parallel on the low-impedance side and in series on the high-impedance side. As you will see, one of the baluns is reversed giving the desired ratio of 6.25:1. Other combinations can produce different fractional-ratios (other than $1{:}n^2$ where n is 1, 2, 3, . . .), like 2.25:1 and 1.78:1. Because very little practical design information is available regarding this family of very broadband baluns,[26,27] this section also includes my high-frequency analysis of the 6.25:1 parallel-type balun. It should also be pointed out that very little sacrifice in performance occurs whether the load is grounded at its center or at the bottom (as an unun).

Figure 5-3 shows the coiled-wire version of the 6.25:1 parallel-type balun. For analysis purposes, the voltages and currents are also shown. As can be seen, the top 4:1 balun is connected as a step-down balun, while the bottom 4:1 balun is connected as a step-up balun. The baluns are in series on the high-impedance side (on the left) and in parallel on the low-impedance side (on the right). As **Figure 5-3** illustrates, the lower 1:4 balun adds a current of $0.5I_1$ to the load, resulting in a total current of $2.5I_1$. Thus, the impedance transformation ratio is 2.5^2, or 6.25:1.

For maximum high-frequency response, each transmission line should see a load equal to its characteristic impedance. In other words, they should be "flat" lines. If the high side on the left is 50 ohms, then 40 ohms appears on the input of the top balun and 10 ohms on the input of the bottom balun. Consequently, the optimum characteristic impedance for *all* transmission lines is 20 ohms. On the low-impedance side on the right, the top balun wants to see 10 ohms, while the bottom balun wants to see 40 ohms. Because 10 ohms in parallel with 40 ohms equals 8 ohms, each balun conveniently sees its ideal load and a broadband ratio of 6.25:1 is obtained.

If the balun is required to match 50-ohm cable (on the right side) to a balanced load of 312.5 ohms (on the left side), the same analysis shows that the optimum characteristic impedance of all the transmission lines is 125 ohms.

Because the parallel-type balun (or unun) sums

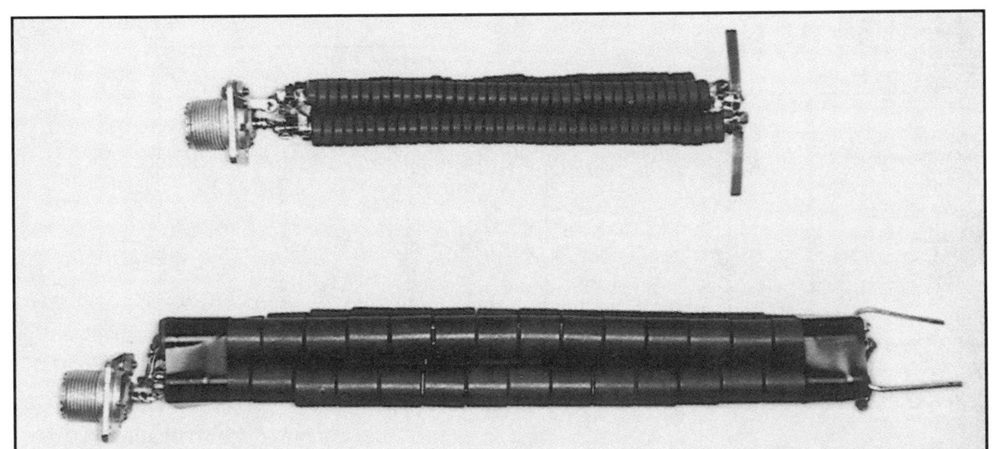

Photo 5-C. Two beaded-line versions of the parallel-type 6.25:1 balun (and unun). The top transformer is designed to match 50-ohm cable to 8 ohms. The bottom transformer is designed to match 50-ohm cable to 312.5 ohms.

voltages of equal delays and therefore has no built-in high-frequency cut-off, it has a real advantage over the series-type balun on the VHF bands and above. Furthermore, beaded transmission lines with low-permeability ferrite beads (125 and less) can be used, resulting in high efficiencies. On the HF band, where coiled windings are generally used, the series-type balun is preferred because of its simplicity.

Photo 5-C shows two beaded-line 6.25:1 transformers. The top balun, designed to match 50-ohm cable (on the left) to 8 ohms (on the right), uses low-impedance coaxial cable lines. The schematic diagram is shown in **Figure 5-4**. It has 5 inches of 0.375 inch OD beads (permeability 125) on four coaxial cables with characteristic impedances of 20 ohms. The inner-conductors of No. 12 H Thermaleze wire have two layers of Scotch No. 92 polyimide tape. The outer braids, from small coaxial cable or 1/8-inch tubular braid, are also wrapped tightly with the same tape in order to preserve the 20-ohm characteristic impedance. When matching 50 ohms to 8 ohms, the response is essentially flat from 10 MHz to beyond 100 MHz (the limit of my simple bridge). Under this matched condition, this balun can easily handle the full legal limit of amateur radio power. Furthermore, it has practically the same performance when operating as an unun (both terminals 1 and 2 grounded). In the unun application, the bottom transmission line has no voltage along it and, therefore, requires no beads.

The bottom balun in **Photo 5-C**, which is designed to match 50-ohm cable to a balanced load of 312.5 ohms, has 8 inches of 0.5 inch OD beads on 125-ohm twin-lead transmission lines. The ferrite beads also have a permeability of 125. The wires are No. 14 H Thermaleze wire and are covered with Teflon sleeving. They are further separated by No. 18 Teflon tubing. When matching 50-ohm cable (on the right side) to 312.5 ohms (on the left side), the response is essentially flat from 20 MHz to over 100 MHz. Under this matched condition, this balun can also easily handle the full legal limit of amateur radio power. Additionally, this transformer performs practically as well when used as an unun.

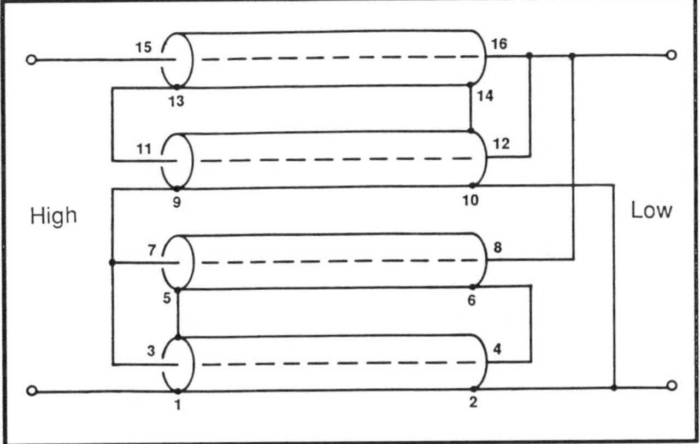

Figure 5-4. Schematic diagram of the coaxial-cable version of the parallel-type 6.25:1 balun (and unun).

Sec 5.3 9:1 Baluns

The broadband 9:1 balun, matching 50-ohm cable to a balanced load of 450 ohms, is one of the most difficult ones to construct because high-impedance transmission lines (150 ohms) are required for maximum high-frequency response, and greater reactances are needed in order to isolate the input from the output. So one can appreciate the task at hand, this section also provides a brief review of the theory of these devices.[2]

Figure 5-5 shows the high- and the low-frequency models of the Guanella 9:1 balun that connect three transmission lines in series at the high-impedance side and in parallel at the low-impedance side. Because

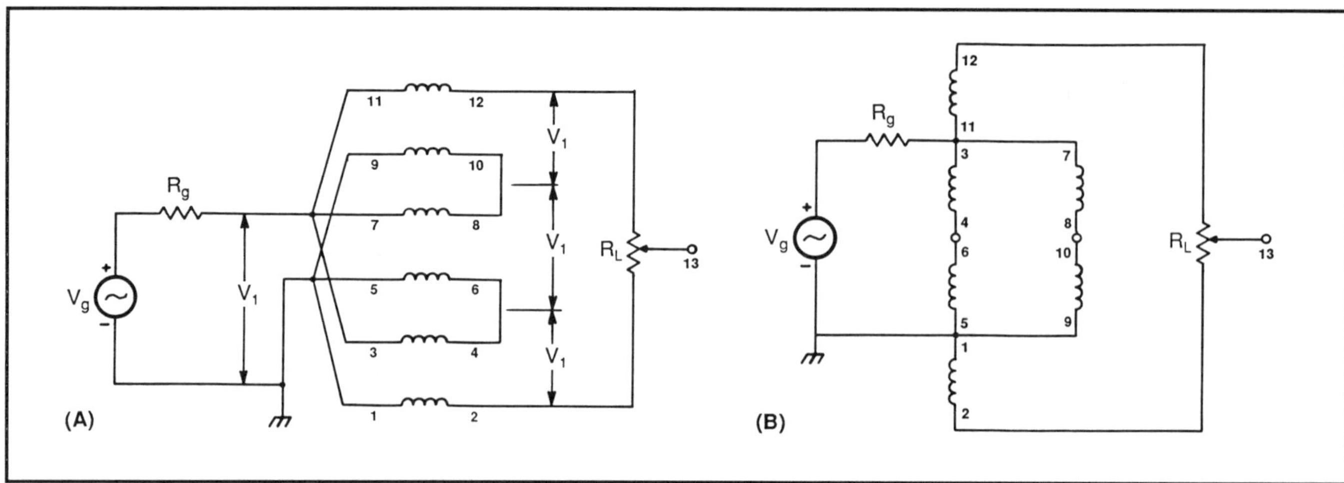

Figure 5-5. Models of Guanella's 1:9 balun. The high-frequency model, (A), assumes that $Z_o = R_L/3$, and therefore V_2, the output of each transmission line, equals V_1. The low-frequency model, (B), assumes no energy is transmitted to the load, R_L, by a transmission line mode.

Guanella's baluns (which can be easily converted to ununs) sum voltages of equal delays, they offer the highest frequency capability.

The high-frequency model (**Figure 5-5A**) assumes there is sufficient choking reactance in the coiled (or beaded) transmission lines to isolate the input from the output and only allow transmission line currents to flow. Under this condition, the analysis is rather straightforward as it only involves transmission line theory. Simply stated—the maximum high-frequency response occurs when each transmission line is terminated in a load equal to its characteristic impedance, Z_0. Thus, the transmission lines in the 9:1 balun have no standing waves. Because each transmission line sees one third of the load, the optimum value of Z_0 is $R_L/3$. Except for parasitics in the interconnections and self-resonances in coiled windings, Guanella's approach is literally "frequency independent."

On the other hand, the low-frequency analysis of the Guanella 9:1 balun is most important because it reveals the major difficulty in designing them for low-loss, wideband operation. **Figure 5-5B** is the model for determining the low-frequency response. It assumes that no energy is transmitted to the load by a transmission line mode. Although the terminology and analysis is the same as that used for conventional autotransformers, the similarity ends when there is sufficient choking reactance to only allow for the efficient transmission line mode.

As with conventional transformers, one can analyze the low-frequency response of the 9:1 balun from either the low or high-impedance side. By putting the generator on the low-impedance side in **Figure 5-5B**, I've chosen to analyze it from that side. With the output open-circuited, the generator sees four coiled (or beaded) lines connected in series-parallel. The net result is that the generator sees the reactance of only *one* coiled (or beaded) line. To prevent a shunting current to ground (and/or autotransformer operation), the reactance the generator sees should be much greater than R_g (at least by a factor of 10 at the lowest frequency of interest). The inductance of the coiled or beaded line that prevents the unwanted currents is still known as the magnetizing inductance, L_M.

What's important to note here is that the low-frequency model of the Guanella 4:1 balun does not have the series-parallel combination of coiled or beaded lines.[2] Only two lines, which are in series, exist in its model. Therefore, for a two-core Guanella 4:1 balun having the same number of turns (and same cores) as a 9:1 Guanella balun, its low-frequency response is better by a factor of two!

Another advantage that goes to the Guanella 4:1 balun when matching 50 to 200 ohms, is in the number of turns that can be wound on the same cores. Since 4:1 baluns require characteristic impedances of 100 ohms (instead of 150), the width of the transmission lines is considerably less, thus allowing for more turns. Also, as will be shown later, the efficiency of the 4:1 balun is greater because the potential drops along the transmission lines are lower (less dielectric loss). Finally, as you can see from **Figure 5-5B**, by also grounding terminal 2 (unun operation), windings 1-2 and 3-4 are both shorted—degrading the low-frequency response because L_M is reduced by one third.

Another interesting analysis with baluns and ununs concerns the potential gradients (voltage drops) along the transmission lines. Because the loss with these transformers, when transferring the energy via a transmission line mode, is a dielectric-type (voltage dependent), the higher the gradient, the greater the loss. The interesting cases occur when the load is: a) floating, b) grounded at the center, and c) grounded at the bottom (an unun).

Floating load. With terminal 13 in **Figure 5-5A** ungrounded, the top transmission line has a gradient of $+V_1$ and the bottom transmission line has a gradient of $-V_1$. The center transmission line has a gradient of zero. Therefore, the center transmission line only acts as a delay line and doesn't require a magnetic core or beads. As a result, the top and bottom cores (or beads) account for the dielectric loss.

Load grounded at the center. With terminal 13 grounded at the center of the load, the top transmission line has a gradient of $+V_1$, the bottom transmission line has a gradient of $-V_1$, and the center transmission line has a gradient of $-V_1/2$. This configuration results in about 25 percent more loss because of the extra gradient along the center transmission line. Incidentally, this condition exists when matching into balanced systems like 450-ohm transmission lines or antennas because they have virtual grounds at the center of the loads they present.

Load grounded at the bottom. With terminal 13 grounded at the bottom of the load (an unun), the top and center transmission lines have gradients of $+V_1$. The bottom transmission line has no gradient, and therefore no loss. It only acts as a delay line and thus requires no magnetic core or beads.

Sec 5.3.1 Some Practical 9:1 Balun Designs

Photo 5-D shows three versions of the broadband Guanella 9:1 balun designed to match 50-ohm cable to 450-ohm loads. The transformer on the left has 15 bifilar turns of No. 18 hook-up wire on each of the three ferrite toroids with a 2.4-inch OD and permeability of 250. The wires are further separated by No. 16 Teflon tubing, resulting in a characteristic impedance close to 150 ohms (the optimum value). The cores in this balun, as well as the other two that follow, are spaced 1/4 inch apart by sections of acrylic. In matching 50 to 450 ohms, the response is essentially flat from 1.7 to 45 MHz. In this matched condition, this transformer can easily handle 500 watts of continuous power and 1 kW of peak power.

The transformer in the center of **Photo 5-D** has 14 bifilar turns of No. 16 SF Formvar wire on each of the three ferrite toroids with a 2.4-inch OD and permeability of 250. The wires are covered with Teflon sleeving and further separated by No. 16 Teflon tubing. The characteristic impedance is also close to the optimum value of 150 ohms. In matching 50 to 450 ohms, the response is essentially flat from 3.5 to 45 MHz. In this matched condition, this transformer can easily handle 1 kW of continuous power and 2 kW of peak power. **Photo 5-E** shows this balun mounted in a minibox 6 inches long by 5 inches wide by 4 inches high.

The transformer on the right in **Photo 5-D** is designed to handle 1 kW of continuous power and 2 kW of peak power from 1.7 to 45 MHz. Because this transformer uses larger cores with slightly higher permeabilities (2.68-inch OD and 290 permeability), which are not as popular as those used in the other

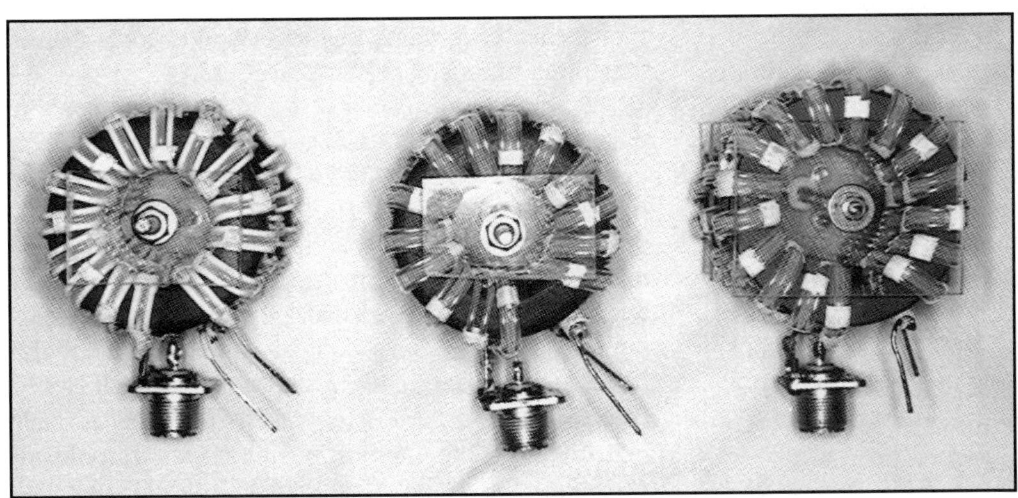

Photo 5-D. Three broadband Guanella 9:1 baluns designed to match 50-ohm cable to 450 ohms. The transformer on the left, using No. 18 hook-up wire, can handle 500 watts from 1.7 to 45 MHz. The transformer in the center, using No. 16 wire, can handle 1 kW from 3.5 to 45 MHz. The transformer on the right (with larger cores), using No. 16 wire, can handle 1 kW from 1.7 to 45 MHz.

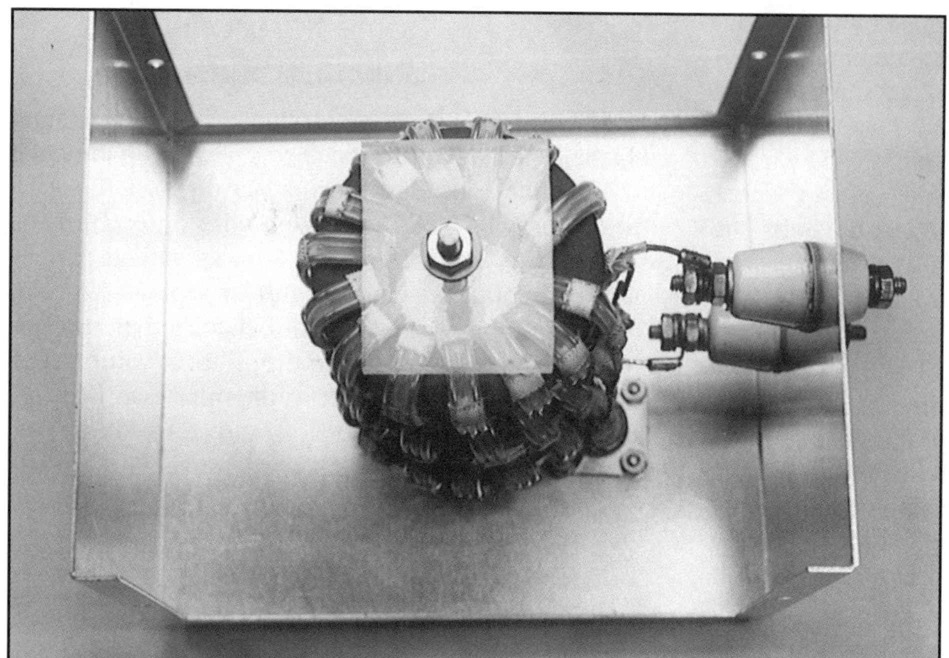

Photo 5-E. The Guanella 9:1 balun mounted in a large minibox.

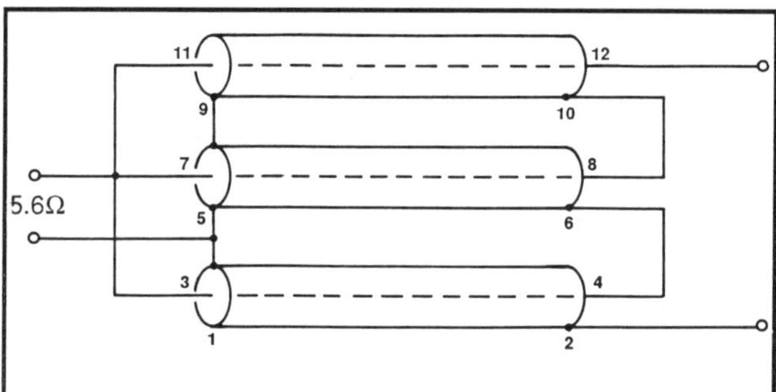

Figure 5-6. Schematic diagram of a coaxial cable (beaded or coiled) 9:1 Guanella balun (or unun). This design is especially useful in matching 50-ohm cable to 5.6 ohms over a very wide bandwidth.

two baluns, it is much more expensive to construct. It has 16 bifilar turns of No. 16 SF Formvar wire on each toroid. The wires are covered with no. 16 Teflon sleeving and further separated with No. 16 Teflon tubing. In matching 50 to 450 ohms, the response is flat from 1.7 to 45 MHz.

Finally, **Figure 5-6** shows the schematic diagram of a 9:1 balun (or unun) using coaxial cables wound around ferrite cores or threaded through ferrite beads. This form of the transformer is especially useful when matching 50-ohm cable to 5.6 ohms because the choking reactance of the magnetizing inductance, L_M, only need be much greater than 5.6 ohms. **Photo 5-F** shows two different designs.

The transformer on the top has 9-1/2 turns of low-impedance coaxial cable on each rod. The rods are 1/2 inch in diameter, 2-1/2 inches long, and have a permeability of 125. The low-frequency response of this balun is quite insensitive to the length and permeability of the rods.[2] The inner conductors of the coaxial cables are No. 12 H Thermaleze wire with two layers of Scotch No. 92 tape. The outer braids are made from small coax (or 1/8-inch tubular braid). They are further wrapped with Scotch No. 92 tape to preserve the low-impedance of 17 ohms. In matching 50-ohm cable to 5.6 ohms, the response is essentially flat from 1.7 to 30 MHz. The optimum impedance level was found when matching 40 to 4.45 ohms. The addition of another layer of Scotch No. 92 tape would optimize this transformer at the 50:56-ohm level, and the high-frequency response would exceed 100 MHz. This transformer is very efficient and should handle the full legal limit of amateur radio power easily.

The bottom transformer in **Photo 5-F** is designed to match 50-ohm cable to a load of 5.6 ohms in the VHF band. It uses 3-1/2 inches of beads on three low-impedance coaxial cables. The ferrite beads have an OD of 3/8 inches and a permeability of 125. The inner conductors of the coaxes are No. 12 H Thermaleze wire with one layer of Scotch No. 92 tape. The outer conductors are from small coaxes or 1/8-inch tubular braid, and are also tightly wrapped with Scotch No. 92 tape to preserve the low characteristic impedance. In matching 50-ohm cable to 5.6 ohms, the response is essentially flat from 7 MHz to over 100 MHz (the limit of my test equipment). This 9:1 balun (which can be used as an unun) can handle the full legal limit of amateur radio power under matched conditions, because of the low permeability beads and the low voltage gradients along the lengths of its transmission lines.

Sec 5.4 Concluding Remarks on 6:1 and 9:1 Baluns

One of the most important properties of broadband baluns and ununs (which all use ferrites) is their capability of having extremely high efficiencies. Knowing the loss mechanism in these transformers and the tradeoff in low-frequency response for efficiency, allows one to optimize their applications. In the paragraphs that follow, I'll discuss the losses and trade-offs involved with the transformers presented in the preceding sections. The approach used here should be applicable to all forms of transmission line transformers. This section ends with a review of two recent articles that contained 9:1 baluns. As you'll see, I have some rather different views on the claims made in these articles.

Accurate measurements on many broadband ununs have found the losses to be related to the permeability and the impedance level.[2] Permeabilities greater than 300 resulted in excessive losses. Because these losses are unlike the conventional transformer whose losses are current-dependent, it can only be assumed that their losses are voltage dependent; in other words a dielectric-type loss. Therefore, higher-impedance transformers have higher voltage gradients along their transmission line and, thus, have greater losses. Additionally, it was found that the higher the permeability, the greater the loss with frequency. Taking into account the accurate measurements and the factors noted above, I offer these loss values for the transformers in the preceding sections:

1.56:1 Ununs. The 1.56:1 ununs (either step-up or step-down) used in series with Guanella 4:1 baluns to form 6.25:1 baluns have the lowest potential gradients along their transmission lines. Voltage drops of only about 0.2 V_1, where V_1 is the input voltage, exist along their transmission lines. Accurate measurements have shown losses, in a matched condition, of only 0.04 dB. If the cores, which have a permeability of 250, were replaced with cores having a permeability of 125, the losses could be as low as 0.02 dB over much of the passband. The low-frequency response would still be acceptable at 1.7 MHz. This unun is a natural for matching into 75-ohm hard line when long transmission lines are required.

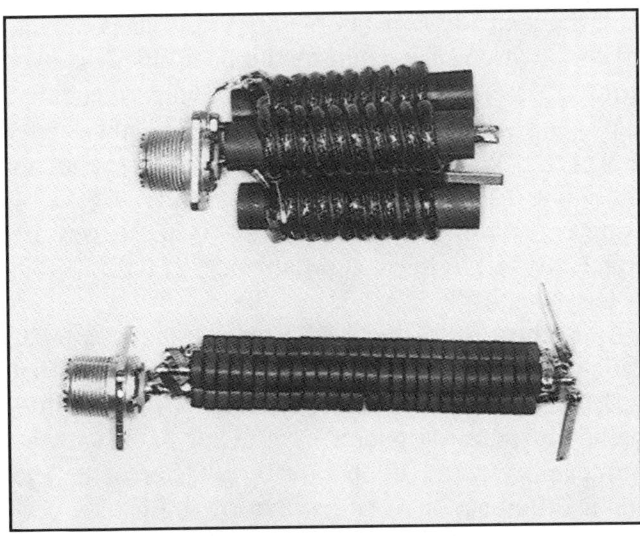

Photo 5-F. Two versions of the coaxial-cable 9:1 Guanella balun (or unun) designed to match 50-ohm cable to 5.6 ohms.

6.25:1 and 9:1 Low-impedance Baluns. Baluns matching 50-ohm cable to 8 or 5.6 ohms, also have very low voltage drops along their transmission lines. Generally, they are about twice that of the 1.56:1 unun. Therefore, the losses with these baluns should be on the order of 0.1 dB in their passbands.

6.25:1 High-impedance Baluns. The losses in the series-type baluns are mainly in the 1:4 Guanella baluns, which have potential gradients of about 1.25 V_1, where V_1 is the input voltage. From previous measurements at this impedance level, the suggestion is that the losses (with ferrites of 250 permeability) should be about 0.1 dB at 7 MHz and 0.2 dB at 30 MHz. By using toroids with permeabilities of 125, the losses could be 0.07 dB and 0.15 dB, respectively. However, with a permeability of 40, the losses could be as low as 0.05 dB within the passband. However, one must consider the sacrifice in low-frequency response incurred when using these lower-permeability ferrites. With a permeability of 125, it's poorer by a factor of 2. With a permeability of 40, it's poorer by a factor of 8! The 6.25:1 parallel-type balun in this article uses ferrite beads with a permeability of 125 and, therefore, should have similar losses to its series-type counterpart.

9:1 High-impedance Baluns. As was shown in the preceding section, the potential gradient along two of the transmission lines is V_1, where V_1 is the input voltage. The third transmission line, with a balanced load (or as an unun), has no potential gradient and, consequently, no loss in its core. Because the loss with the series-type balun mainly exists in one core, the loss with the 1:9 balun should be a little less than twice as great. With ferrite cores of 250 permeability, the suggested losses are 0.2 dB at 7 MHz and 0.4 dB at 30 MHz. With cores of 125 permeability, the losses are about 0.14 and 0.28 dB, respectively. Again by using cores with permeabilities of 40, the losses are practically negligible—approximately 0.1 dB within its passband.

As in the case of the 1:6.25 baluns above, similar trade-offs occur in the low-frequency response. That is, if 125 permeability cores are used, the low-frequency response is poorer by a factor of 2; with 40 permeability cores, it's poorer by a factor of 8. The major difference here is that the low-frequency performance of the 9:1 balun, as seen by its low-frequency model, isn't as good as the 4:1 balun that controls the low-frequency response of the series-type 1:6.25 balun. Additionally, it should be pointed out that all of the suggested losses for the transformers in this chapter are for matched conditions—that is with VSWRs of 1:1. If the VSWR is 2:1 due to a load twice as large as the objective, the input voltage to the balun increases by about 40 percent. Therefore, the losses should increase by close to the same percentage.

In closing this section, I would like to report my findings on two recent articles in amateur radio journals that also described 9:1 baluns matching 50-ohm cable to 450 ohms. One[5] advocated using three 150-ohm coaxial cables threaded through high-permeability ferrite beads—a 9:1 Guanella balun. Because of the low voltage-breakdown capability of the coaxial cable and the high loss found by accurate measurements on ununs using these high-permeability ferrites, the design was suspect. I built a copy of the design and found it to be, as expected, unable to handle any appreciable power. The second article[28] advocated using 14 trifilar turns of "magnet wire" on a 2-inch OD powdered-iron core (permeability of 10). This balun was also constructed and tested. Again, as was expected, when matching 50-ohm cable to a floating load of 450 ohms, the 9:1 balun barely reached a true 9:1 ratio at 7 MHz. Above 7 MHz, the ratio became greater than 9:1 and also introduced a reactive component. Below 7 MHz, there was insufficient choking reactance to prevent flux in the core. My three objections to this design are: 1) a trifilar design has a poor high-frequency response because it sums a direct voltage with a delayed voltage that traverses a single transmission line and a delayed voltage that traverses two transmission lines, 2) the characteristic impedances of the transmission lines are only 50 ohms (the objective is 150 ohms), and 3) the low-frequency response is poor because of the low-permeability powdered-iron core. I do not recommend either of the designs in these two articles.

Sec 5.5 12:1 Baluns

Over the years, the broadband, 12:1 balun has been of special interest to users of rhombic and V antennas. With the aid of this balun, certain advantages over multielement arrays can be fully exploited. Rhombics and Vs are easier to construct both electrically and mechanically, and there are no particularly critical dimensions or adjustments. Furthermore, they give satisfactory gain and directivity over a 2-to-1 frequency range. These antennas have also been found to be more effective in reception. Because their designs can present input impedances of 600 ohms, and very long lengths of highly efficient 600-ohm open-wire line can be used between the shack and the antenna, an efficient and broadband 12:1 balun is a natural for this application. However, to my knowledge, satisfactory baluns have not been available for this use.

I will present two versions of a series-type balun designed to match 50-ohm cable to a balanced load of 600 ohms. One is a high-power unit designed to handle the full legal limit of amateur radio power over a bandwidth of 7 to 30 MHz. The other is a medium-power unit capable of handling approximately one-half the legal limit of amateur radio power from 3.5 to 30 MHz. Both baluns use a 1:1.33 unun in series with a 1:9 Guanella balun.

As you will see, these baluns are not especially easy to design and construct. The major difficulties lie in trying to obtain sufficient choking reactances in the coiled windings to meet the low-frequency requirements, and large enough characteristic impedances of the windings to meet the high-frequency requirements. Because a coiled winding with a characteristic impedance of 200 ohms (the objective) is practically impossible to obtain with any reasonable wire size and number of turns, I used the compensating technique first described in my book.[2] Because the characteristic impedances of the 9:1 Guanella balun are somewhat less than 200 ohms, a compensating effect (and hence higher frequency response) can be obtained by having a higher (than the normal objective) characteristic impedance of the windings in the 1:1.33 unun. Earlier work (also described in my book) presented a 12:1 balun using a 1:3 unun in series with a

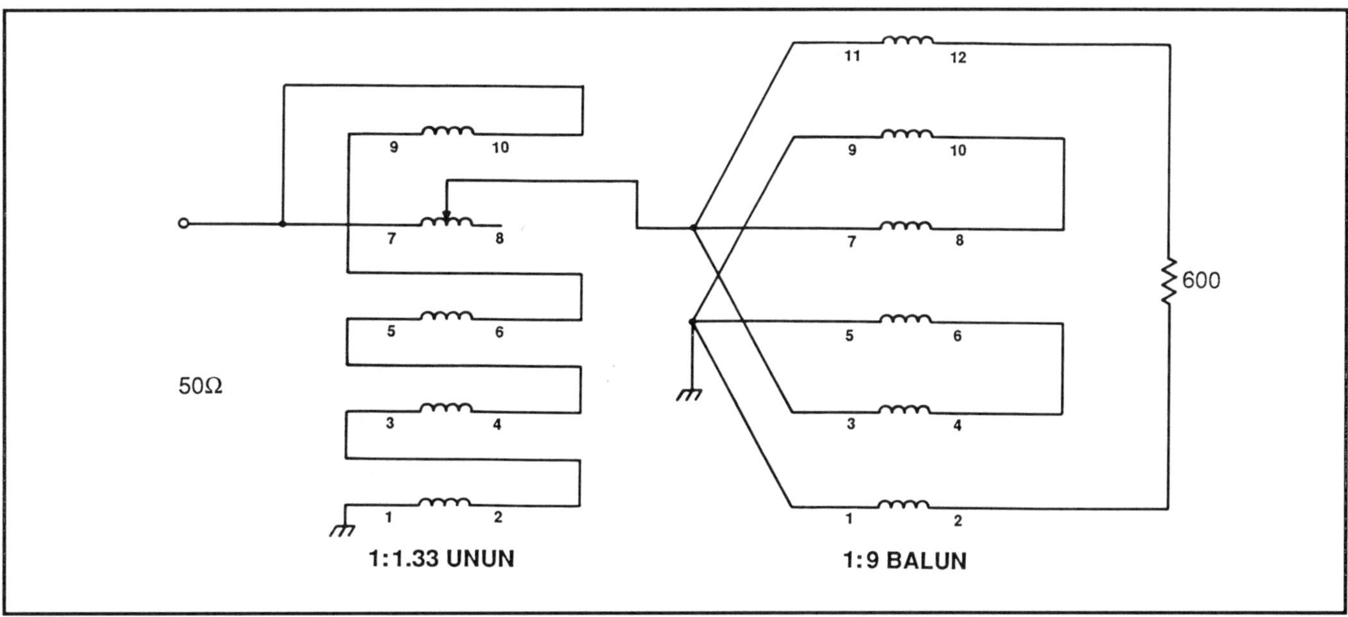

Figure 5-7. Schematic diagram of the series-type 12:1 balun using a 1:1.33 unun in series with a 1:9 Guanella balun.

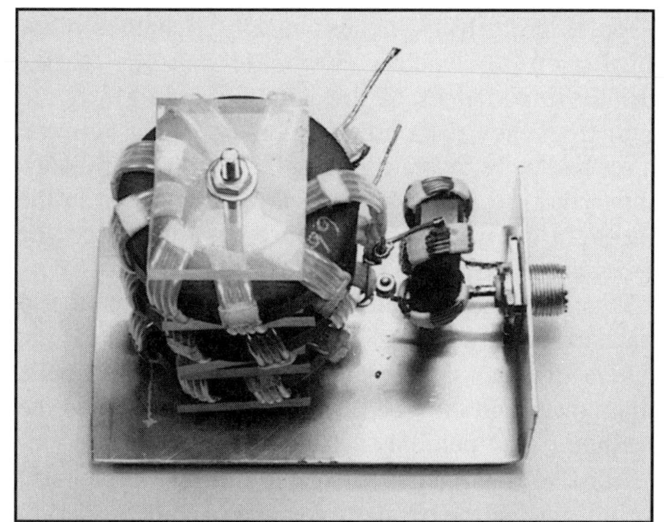

Photo 5-G. Top view of the high-power 12:1 balun.

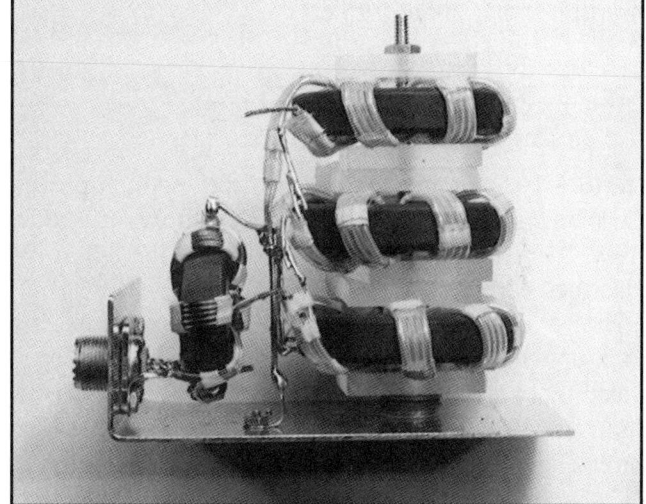

Photo 5-H. Side view of the high-power 12:1 balun.

1:4 Ruthroff balun. The baluns presented in this section using a 1:1.33 unun in series with a Guanella 1:9 balun, are much improved designs.

Sec 5.5.1 A High-power 12:1 Balun

Figure 5-7 shows the schematic diagram of the series-type 12:1 balun used in both the high and medium-power versions. **Photo 5-G** shows a top view of the high-power balun. **Photo 5-H** shows a side view.

The 1:1.33 unun has 5 quintufilar turns on a 1.5-inch OD ferrite toroid with a permeability of 250. Winding 7-8 is No. 14 H Thermaleze wire and the other four are No. 16 H Thermaleze wire. Winding 7-8 is also tapped at 3 turns from terminal 7.

The 1:9 Guanella balun has 8 bifilar turns of tinned No. 16 wire on each of the three toroids. Each wire is covered with Teflon tubing and further separated by two Teflon tubings. The characteristic impedance of the windings is about 190 ohms (the objective is 200 ohms). The ferrite toroids have an OD of 2.4 inches and a permeability of 250. The spacing between the toroids is 1/2 inch.

In matching 50-ohm cable to a balanced load of 600

Photo 5-I. Top view of the medium-power 12:1 balun.

Photo 5-J. Side view of the medium-power 12:1 balun.

ohms, the response is literally flat (within a percent or two) from 7 to 30 MHz. Within this bandwidth, it is capable of handling the full legal limit of amateur radio power. In a matched condition, the expected insertion loss is about 0.25 dB.

Sec 5.5.2 A Medium-power 12:1 Balun

Photo 5-I shows the top view of the medium-power 12:1 series-type balun. **Photo 5-J** shows the side view. This balun also has the same 1:1.33 unun described earlier.

The 1:9 Guanella balun has 11 bifilar turns of No. 18 hook-up wire on each toroid. The wires are further separated by two No. 18 Teflon tubings. The characteristic impedance of the windings is about 170 ohms. The toroids have an OD of 2.4 inches and a permeability of 250. The spacing between the toroids is also 1/2 inch.

In matching 50-ohm cable to a balanced load of 600 ohms, the response varies less than 5 percent from 3.5 to 30 MHz. Within this bandwidth, the balun can handle about one-half the legal limit of amateur radio power. As with the high-power version, the expected insertion loss is also 0.25 dB.

Sec 5.6 Concluding Remarks on 12:1 Baluns

Many of the concluding remarks from the discussion on 6:1 and 9:1 baluns (Sec 5.4) also apply to 12:1 baluns; therefore, I won't repeat them here. But the following four remarks are specific to 12:1 baluns and warrant mentioning:

First, high-impedance transmission line transformers like the 12:1 balun are particularly sensitive to metallic enclosures. If a minibox were to be used, I would suggest the one shown in **Photo 5-E**, which is 6 inches long by 5 inches wide by 4 inches high. Smaller metallic enclosures would reduce the characteristic impedances of the windings and affect the high frequency response. Even the spacing between cores had to be increased from 1/4 inch (for a 50:450-ohm balun) to 1/2 inch. The subchassis shown in the photographs were used because they provided the necessary electrical and mechanical support.

Second, the 12:1 baluns described in this section also make excellent ununs, albeit with some compromise in the low-frequency response. I would suggest using the high-power unit only between 14 and 30 MHz and the medium-power unit only between 7 and 30 MHz.

Third, for the readers interested in VHF operation, I would suggest the parallel-type approach described in the earlier section on 6:1 baluns. In this case, a 9:1 Guanella balun is connected in series-parallel with a 1:4 Guanella balun. This would produce a broadband ratio of 12.25:1. By using 170-ohm twin-lead (about 10 inches long) threaded through ferrite beads with a permeability of 125, it appears that it is possible to match 50-ohm cable to a balanced load of 612.5 ohms throughout the VHF band.

Fourth, by using torids with a permeability of 125 in the 1:9 Guanella baluns (of the 12:1 baluns), the insertion loss would be reduced by around one half (0.12 dB), with a trade-off in low-frequency response. The high-power unit would now cover about 10 to 30 MHz, and the medium-power balun would cover about 7 to 30 MHz.

Chapter 6

The 4:1 Unun

Sec 6.1 Introduction

From an analysis standpoint, the 4:1 unun can be said to have received the most attention in the literature. It began with Ruthroff's introduction and complete analysis of this device in his classic paper published in 1959.[9] Ruthroff's paper then became the industry standard for this class of devices known as *transmission line transformers*. These are devices that transmit the energy from the input to the output by an efficient transmission line mode, and not by flux linkages (as in conventional transformers).

However fifteen years earlier, Guanella had introduced, in his classic 1944 paper,[3] the first broadband baluns by combining coiled transmission lines in a series-parallel arrangement, yielding ratios of $1:n^2$ where n = 1, 2, 3, and so on. Recently, it was shown that Guanella's technique also lends itself to ununs as well.[2] In fact in this chapter, you will see that his technique of summing voltages of equal delays promises to yield high-power designs capable of operating on the VHF and UHF bands.

The 4:1 unun also exemplifies (more than any other transformer) the many choices that can be made in its design. These include: 1) Ruthroff's or Guanella's designs, 2) wire or coaxial cable transmission lines, 3) coiled or beaded lines, 4) rods or toroids, 5) low-power or high-power designs, 6) HF, VHF, or UHF designs, and 7) the trade-offs in efficiency for low-frequency response or for high VSWR. The 4:1 unun is the most prevalent of all the ununs. It finds extensive use in solid-state circuits and in many antenna applications involving the matching of ground-fed antennas—where impedances of 12 to 13 ohms must be matched to 50-ohm coaxial cable. This chapter provides information on many 4:1 unun designs.

Sec 6.2 The Ruthroff 4:1 Unun

Figure 6-1 illustrates two versions of Ruthroff's approach to obtaining a 4:1 unbalanced-to-unbalanced transformer (unun). As can be seen, one uses a coiled wire transmission line, while the other uses a coiled coaxial cable. Depending upon the frequency, beaded transmission lines may also be used.

Ruthroff's design uses a single transmission line connected in, what I call, the "bootstrap" configuration. That is, terminal 2 is connected to terminal 3,

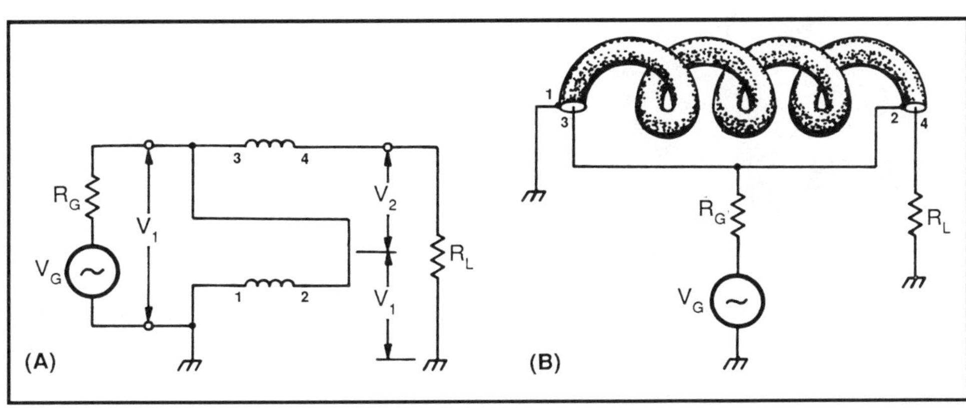

Figure 6-1. The Ruthroff 4:1 unun ($R_L=4R_G$): (A) coiled bifilar winding; (B) coiled coaxial cable.

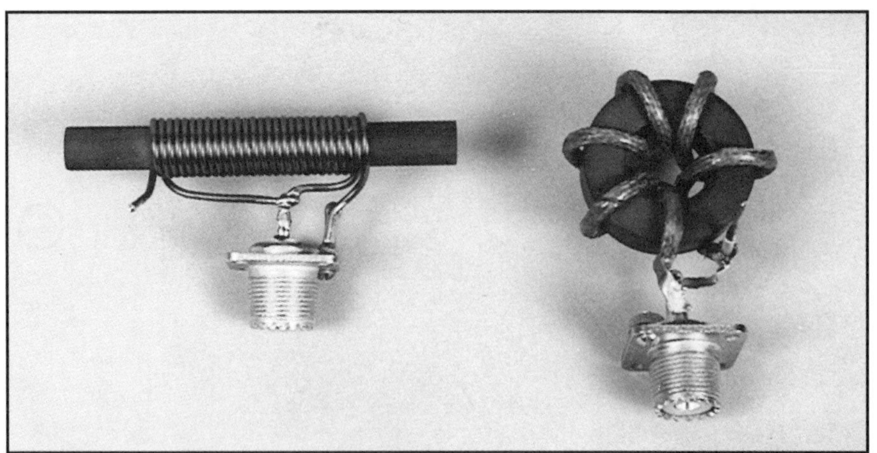

Photo 6-A. Two versions of the Ruthroff 4:1 (50:12.5 ohm) unun: coiled wire rod (on the left); coiled coaxial cable toroid (on the right).

lifting the transmission line (at the high-impedance side) by the voltage V_1. If the reactance of the coiled winding or beaded line is much greater than R_G, then only flux-canceling transmission line currents are allowed to flow. It is also apparent that the output voltage is the sum of a direct voltage, V_1, and a delayed voltage, V_2, which traverses a single transmission line. This delay in V_2 eventually limits the high-frequency response. For example, if the electrical length of the line is a half wave, the output is zero. Ruthroff also found that the optimum value of the characteristic impedance of the transmission line (for maximum high-frequency response) is $R_L/2$.

Therefore, the electrical length and characteristic impedance of the transmission line play major roles in Ruthroff's design. Because his work was mainly concerned with small-signal applications, Ruthroff was able to obtain broad bands of a few tens of kilohertz to over a thousand megahertz. This was possible because he used a few turns (5 to 10) of fine wire (Nos. 37 and 38) on high-permeability toroids as small as 0.08 inches in OD. As a result, the phase-delay with these very short transmission lines was very small. However, large-signal (power) applications present an entirely different picture. For operation in the HF band (including 160 meters), transmission lines vary between one to three feet in length (depending upon impedance level). Consequently, phase-delay can play a major role, as will be seen in the following examples.

Sec 6.2.1 50:12.5-ohm Ununs

Photo 6-A shows two examples of efficient, broadband 4:1 ununs matching 50 to 12.5 ohms. The rod version (on the left) has 14 bifilar turns of No. 14 H Thermaleze wire on a low-permeability (125) ferrite rod 0.375 inches in diameter and 3.5 inches long. The connections are shown in **Figure 6-1A**. The cable connector is on the low-impedance side. The response is flat from 1.5 to 30 MHz. In a matched condition, this unun can easily handle the full legal limit of amateur radio power. Because a tightly wound rod unun yields a characteristic impedance very close to 25 ohms (the optimum value), this is quite likely the easiest one to construct that covers the above bandwidth.

The toroidal version (on the right in **Photo 6-A**) has

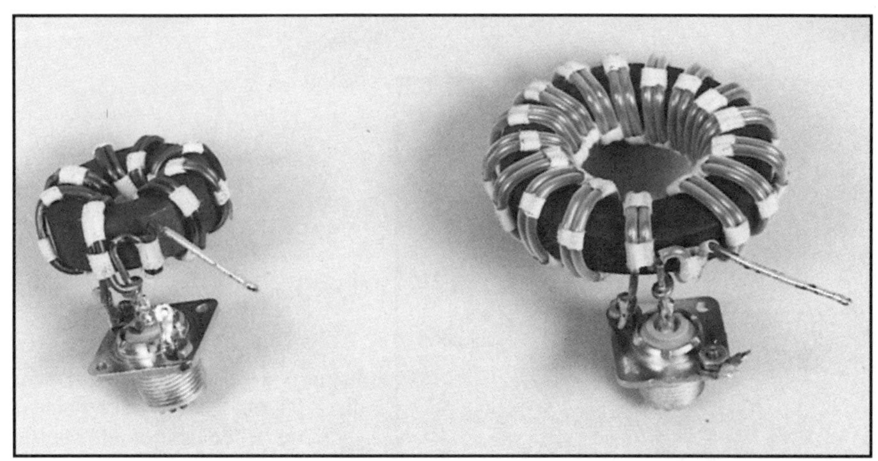

Photo 6-B. Two higher-impedance Ruthroff 4:1 ununs: 100:25 ohm (on the left); 200:50 ohm (on the right).

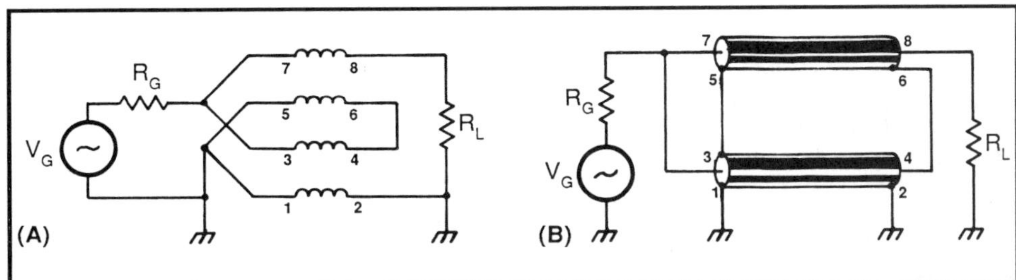

Figure 6-2. The Guanella 4:1 unun (R_L=4R_G): (A) coiled bifilar windings; (B) coiled or beaded coaxial cables.

6 turns of homemade, low-impedance coaxial cable on a 1.5-inch OD ferrite toroid with a permeability of 250. The connections are shown in **Figure 6-1B**. The cable connector is on the low-impedance side. The inner conductor is No. 14 H Thermaleze wire and is covered with Teflon™ tubing. The outer braid is from a small coaxial cable (or from 1/8-inch tubular braid) tightly wrapped with Scotch No. 92 tape in order to obtain the desired characteristic impedance. In matching 50 to 12.5 ohms, the response is flat from 1.5 to 50 MHz. Because the current is evenly distributed on the inner conductor, this small unun has an exceptionally high power capability—at least 5 kW of continuous power and 10 kW of peak power (in a matched condition).

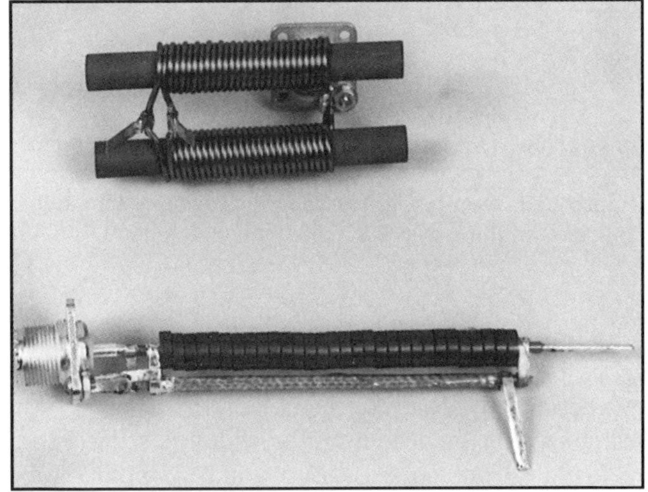

Photo 6-C. Two Guanella 4:1 (50:12.5 ohm) ununs: rod version (on the top), 1.5 to 50 MHz; beaded version (on the bottom), 10 MHz to over 100 MHz.

Sec 6.2.2 100:25-ohm Unun

In some combiner applications, an unun matching 100 to 25 ohms is required. The smaller toroidal version, pictured on the left in **Photo 6-B**, shows a Ruthroff design that can satisfy many of these requirements. It has 8 bifilar turns of No. 14 H Thermaleze wire on a 1.5-inch OD ferrite toroid with a permeability of 250. One wire is also covered with a single layer of Scotch No. 92 tape, providing a characteristic impedance close to the desired value of 50 ohms. In matching 100 to 25 ohms, the response is essentially flat from 1.5 to 30 MHz. This unun can easily handle the full legal limit of amateur radio power.

Sec 6.2.3 200:50-ohm Unun

When dealing with this type of balun, the Ruthroff approach cannot yield the broadband response of the lower-impedance designs shown above. Because more turns are required in order to obtain the necessary choking reactance, and a 100-ohm characteristic impedance that requires more spacing between the wires is used, the cores must be considerably larger. This results in longer transmission lines. Consequently, the high-frequency response is now limited by the greater phase delay of this high-impedance unun.

The larger transformer, shown on the right in **Photo 6-B** is my optimized version of a Ruthroff 200:50-ohm unun. It has 16 bifilar turns of No. 14 H Thermaleze wire on a low-permeability (250) 2.4-inch OD ferrite toroid. Each wire is covered with Teflon tubing, resulting in a characteristic impedance of 97 ohms. Because of the long transmission line (36 inches), the impedance transformation ratio (in matching 200 ohms to 50 ohms) varies from 4 to 4.44 from 1.5 to 30 MHz. A conservative power rating (under a matched condition) is 2 kW of continuous power and 4 kW of peak power. Because this higher-impedance unun has a larger voltage drop along the length of its windings, its loss (a dielectric-type[2]) is a little greater than the lower-impedance ununs described earlier. In a matched condition, the efficiency is about 97 percent, while the others experience efficiencies of 98 to 99 percent.

Sec 6.3 The Guanella 4:1 Unun

Even though Guanella's investigation[3] was directed toward developing a broadband balun to match the balanced output of a 100-watt, push-pull, vacuum-

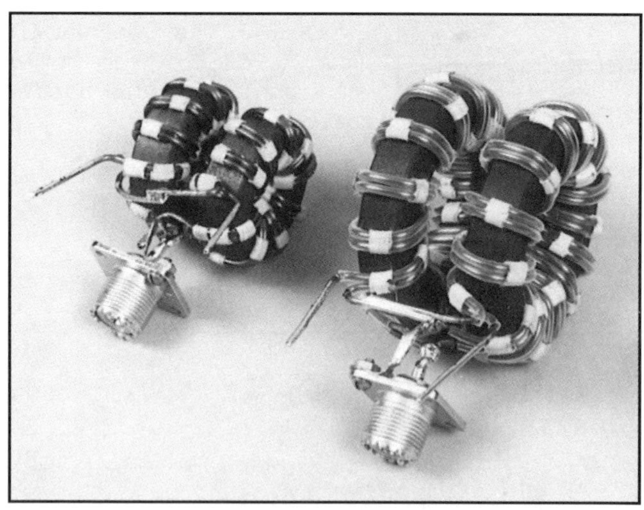

Photo 6-D. Two higher-impedance Guanella 4:1 ununs: 100:25 ohm (on the left); 200:50 ohm (on the right).

tube amplifier to the unbalanced load of a coaxial cable, his technique of connecting transmission lines in a parallel-series arrangement has only recently been recognized as the design for the widest possible bandwidth in an unbalanced-to-unbalanced application.[2] Some have labeled his approach the "equal-delay network".[26] The major difference in Guanella's approach (from Ruthroff's) is that by summing the equal-delay voltages of coiled (or beaded) transmission lines, he minimizes the dependence of the high-frequency response on the lengths of the transmission lines. As was mentioned before, Ruthroff's method of summing a direct voltage with a delayed voltage that traversed a single transmission line has a limited application especially with high-power, high-impedance ununs (like 200:50 and 300:75 ohms).

Furthermore, Guanella's approach is also important in designing high- and low-impedance baluns and ununs with impedance transformation ratios other than 4:1. Connecting three transmission lines in parallel-series results in a 9:1 ratio, four in a 16:1. Also by connecting a fractional-ratio unun in series with his baluns, or by using various combinations of parallel-series transmission lines,[26,27] ununs and baluns are now available with a continuum of ratios from 1.36:1 to 16:1. Moreover, these ratios now make it possible to match 50-ohm cable to impedances as low as 3.125 ohms and as high as 800 ohms. A major factor in the success of these designs rests in the understanding of the low-frequency models of these various transformers.[2] This section looks at the 4:1 unun using Guanella's approach. As in the Ruthroff case, the optimum value of the characteristic impedances of the transmission lines for a Guanella 4:1 transformer is also $R_L/2$.

Sec 6.3.1 50:12.5-ohm Ununs

Figure 6-2 shows the schematic diagrams of the coiled-wire and coaxial cable (coiled or beaded) versions of 4:1 ununs using Guanella's technique of connecting transmission lines in parallel-series arrangements. As can be seen in **Figure 6-2**, the lower transmission lines are grounded at both ends and therefore have no potential drop along their lengths. Thus, the coiling or beading has no effect. The core only acts as a mechanical support and the beads can be removed. In essence, the bottom transmission line plays the important role of a delay line. In addition, the low-frequency response of this form of unun is solely determined by the reactance of the top coiled or beaded transmission line.

The top unun in **Photo 6-C** shows a rod version of Guanella's 4:1 unun. There are 13.5 bifilar turns of No. 14 H Thermaleze wire on low-permeability (125) ferrite rods 0.375 inches in diameter and 3.5 inches long. For ease of connection, one winding is clockwise and the other is counterclockwise. The cable connector is on the high-impedance side. In matching 50 to 12.5 ohms, the response is flat from 1.5 to over 50 MHz! This unun, in a matched condition, is capable of handling the full legal limit of amateur radio power. Furthermore, with the 50-ohm generator on the right (in **Figure 6-2A**) and a 12.5-ohm balanced load on the left (perhaps a Yagi beam), this transformer makes an excellent step-down balun.

The bottom transformer in **Photo 6-C** shows a beaded-coax version of a 50:12.5-ohm step-down unun designed for 2-meter operation. It has 3.5 inches of beaded coax on the top transmission line (**Figure 6-2B**) and no beads on the bottom transmission line. (Actually, the bottom rod in **Figure 6-2A** can also be removed with no change in performance.) The beads are low-permeability (125) ferrite. The inner conductor of the coaxial cable is No. 12 H Thermaleze wire with about 3.5 layers of Scotch No. 92 tape (two 0.5-inch tapes wound edgewise like a window shade), providing a characteristic impedance close to the optimum value. The outer braid is from a small coaxial cable (or from 1/8-inch tubular braid). This home-made coax is further wrapped tightly with Scotch No. 92 tape in order to preserve its low characteristic impedance. The cable connector is on the low-impedance side. The response of this unun is essentially flat from 10 to 100 MHz (the limit of my bridge). It can also (easily) handle the full legal limit of amateur radio power.

Sec 6.3.2 100:25-ohm Unun

The unun on the left in **Photo 6-D** is a Guanella version that matches 100 to 25 ohms. There are 8 bifilar turns of No. 14 H Thermaleze wire on each 1.5-inch OD low-permeability (250) toroid. One toroid is wound clockwise and the other is wound counterclockwise. One of the wires (on each toroid) is covered with one layer of Scotch No. 92 tape. The cable connector is on the low-impedance side. The response is flat from 1.5 MHz to well over 30 MHz. This unun can also handle the full legal limit of amateur radio power.

It is interesting to note that when used as a balun (the ground removed from terminal 2), and placed in series (on the left side) with a 1.78:1 unun (see **Chapter 7**), this compound arrangement provides an excellent balun for matching 50-ohm coaxial cable directly to quad antennas having impedances of 100 to 110 ohms.

Sec 6.3.3 200:50-ohm Unun

The transformer on the right in **Photo 6-D** is an excellent unun (or balun with terminal 2 removed from ground) for matching 50 to 200 ohms. It has 14 bifilar turns of No. 14 H Thermaleze wire on each low-permeability (250) toroid with a 2.4-inch OD. Each wire is covered with Teflon tubing, providing a characteristic impedance of 98 ohms (which is quite good because the optimum value is 100). Again, for ease of connection, one winding is clockwise and the other is counterclockwise. When operating as an unun or a balun and matching 50 to 200 ohms, the response is essentially flat from 1.5 to 30 MHz. A conservative power rating (in a matched condition) is 5 kW of continuous power and 10 kW of peak power. This transformer has been reported to handle peak pulses of 10,000 volts!

Summary

Since its introduction by Ruthroff in 1959,[9] the 4:1 unun has been the most popular transmission line transformer matching unbalanced impedances to unbalanced impedances. As I mentioned at the beginning of this chapter, there are many choices to consider when designing these broadband and efficient transformers. One of the most important choices involves whether to use the Ruthroff or Guanella approach. In fact, the Guanella design should probably be designated a *balun/unun*. Recently, it has become the design of choice in the higher frequency bands. From the designs presented in this chapter, I offer the following recommendations:

1. For ununs in the HF band with impedance levels of 100:25 ohms and lower, the Ruthroff approach is recommended because of its simplicity.
2. For high impedance levels in the HF band (like 200:50 and 300:75 ohms), the Guanella approach is recommended.
3. For low-impedance operation on the VHF band, the beaded-coax Guanella approach is recommended.
4. For high-impedance operation on the VHF band, the coiled-wire Guanella approach appears to be the preferred choice, and should be investigated first. Obviously, the number of turns should be reduced from the examples shown in this chapter because the reactance of the winding is proportional to the frequency.
5. For high-power use on the HF band, the Ruthroff unun with low-impedance coaxial cable on a toroid (on the right on **Photo 6-A**) is recommended. It is easy to construct and can very likely handle more than 5 kW of continuous power.
6. Also, at high-impedance levels, one might consider using lower permeability ferrites for higher efficiencies. Look at permeabilities of 125 and 40.

Chapter 7

1.33:1, 1.5:1 and 2:1 Ununs

Sec 7.1 Introduction

Little practical design information has been available on ununs with impedance transformation ratios of less than 4:1 (these are called *fractional-ratio* ununs). However, many important applications can be found for efficient and broadband ununs with ratios like 1.33:1, 1.5:1, and 2:1. Some examples include the matching of 50-ohm coaxial cable to: a) vertical antennas, inverted Ls, and ground-fed slopers (all over good ground systems), b) 75-ohm hardline cable, c) a junction of two 50-ohm coaxial cables, d) shunt-fed towers performing as vertical antennas, and e) the output of a transceiver or class B linear amplifier when an unfavorable VSWR condition exists.

These three ununs also play an important role in making other useful baluns possible. Examples given in earlier chapters include: a) connecting a 1.5:1 unun (50:75 ohms) in series with a 1:1balun (75:75 ohms) results in a broadband 1.5:1 balun (50:75 ohms); b) connecting a 2:1 unun (50:100 ohms) in series with a 1:1 balun (100:100 ohms) results in a broadband 2:1 balun (50:100 ohms); c) connecting a 1.5:1 unun (50:75 ohms) in series with a 4:1 balun (75:300 ohms) results in a broadband 6:1 balun (50:300 ohms), and d) connecting a 1.33:1 unun (50:66.7 ohms) in series with a 9:1 balun (66.7:600 ohms) results in a broadband 12:1 balun (50:600 ohms).

Recently, it has been shown[2] that a continuum of ratios can now be obtained with ununs matching 50-ohm cable to impedances as low as 3.125 ohms and as high as 800 ohms. In addition, by using higher-order windings (trifilar, quadfilar, etc.), ununs can be constructed with two broadband ratios like 1.5:1 and 3:1, or 2:1 and 4:1. Furthermore, by tapping some of the windings of these higher-order ununs, multimatch transformers can be constructed with many broadband ratios. As a result of this class of fractional-ratio ununs, a continuum of ununs and baluns is now available to match 50 ohms unbalanced to unbalanced or balanced impedances as low as 3.125 ohms and as high as 800 ohms.

My first attempt to obtain ratios less than 4:1 was made by tapping one of the wires in a Ruthroff 4:1 bifilar unun. My experiment met with only moderate success.[2] An adequate low-frequency response with a 1.33:1 ratio was difficult to obtain. Also, the 2:1 ratio

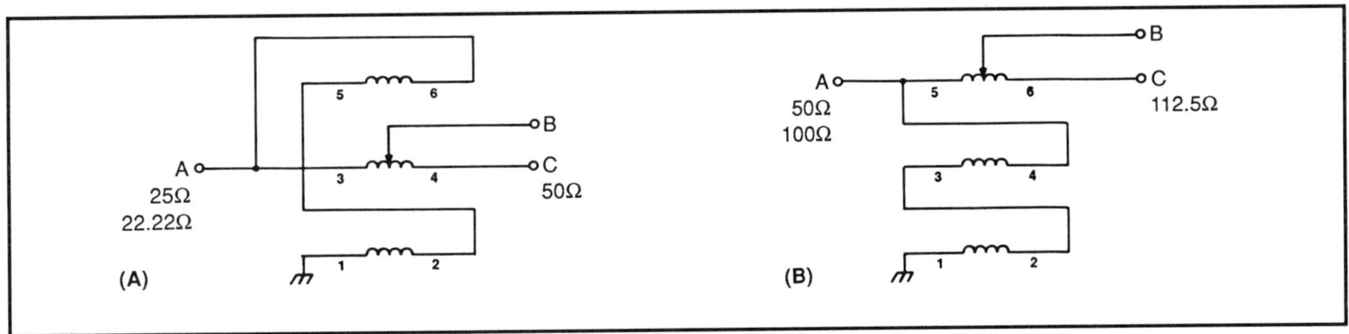

Figure 7-1. Schematic diagrams: (A) matching 50 to 25 ohms (B-A) and 50 to 22.22 ohms (C-A); (B) matching 100 to 50 ohms (B-A) and 112.5 to 50 ohms (C-A).

Photo 7-A. Bottom view of the 2:1 unun designed to match 50 ohms to 25 ohms or 22.2 ohms (Figure 7-1A). The connector is on the low-impedance side.

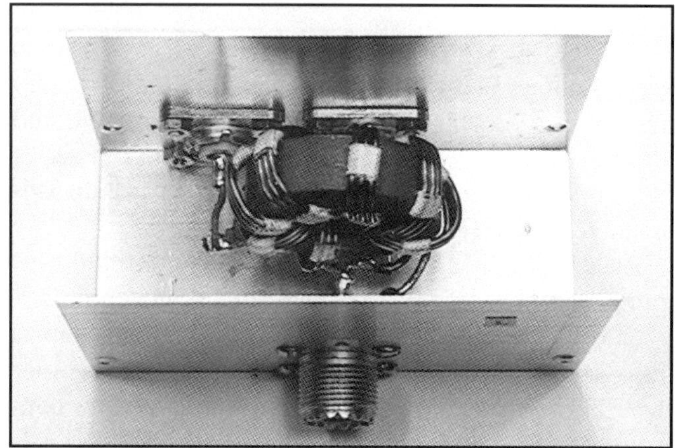

Photo 7-B. The 2:1 unun mounted in a 4-inch long by 2-inch wide by 2.75-inch high minibox.

had considerably greater loss than higher or lower ratios. Recently, I found that higher-order windings (trifilar, quadrifilar, etc.), some with taps, provide much wider bandwidths and higher efficiencies. This chapter describes fractional-ratio ununs using these higher-order windings.

The next section discusses the practical aspects of the 2:1 unun. This unun is not only one of the more useful transformers, but it also serves as a good introduction to the trifilar and quadrifilar designs. What follows is an introduction to the most difficult fractional-ratio unun—the quintifilar design, which results in very broadband 1.33:1 and 1.5:1 ununs. For a more in-depth discussion of these fractional-ratio ununs, I direct interested readers to the second edition of my book on transmission line transformers.[2]

This chapter closes with construction tips. As you will see, these ununs can be difficult to construct.

Sec 7.2 2:1 Ununs

Figure 7-1A shows the schematic diagram of an unun designed to match 50-ohm cable to an unbalanced load of 25 ohms (2:1 ratio with connections A-B) or 22.22 ohms (2.25:1 ratio with connections A-C). It has 6 trifilar turns of No. 14 H Thermaleze wire on a 1.5-inch OD ferrite toroid with a permeability of 250. Winding 3-4 is tapped at 5 turns from terminal 3. **Photo 7-A** is a photograph showing the various connections. The connector is on the low-impedance side. **Photo 7-B** shows the transformer mounted in a CU-3015A (4 inches long by 2 inches wide by 2.75 inches high) minibox. In matching 50 ohms to either 25 or 22.22 ohms, the transformation ratio is constant from 1 to 30 MHz.

Because the transmission lines are very short, this unun does quite well as a step-up transformer. That is, when matching 50 ohms (on the left side) to 100 ohms (connections A-B) or 112.5 ohms (connections A-C) on the right side, the transformation ratio is constant from 1 to 15 MHz. Because of the extremely high efficiency of this transformer (98 to 99 percent under matched conditions), this small version can easily handle the full legal limit of amateur radio power.

Figure 7-1B shows the schematic diagram of an unun designed to match 50-ohm cable to an unbalanced load of 100 ohms (2:1 ratio with connections A-B) or 112.5-ohms (2.25:1 ratio with connections A-C). It has 7 trifilar turns on a 1.5-inch OD ferrite toriod with a permeability of 250. The top winding 5-6 is No. 14 Thermaleze wire and is tapped at 6 turns from terminal 5. The other two windings are No. 16 H Thermaleze wire. **Photo 7-C** shows the various connections. The connector is on the low-impedance side. In matching 50-ohm cable to 100 ohms (A-B) or 112.5 ohms (A-C), the transformation ratio is constant from 1 to 30 MHz.

Again, because the transmission lines are very short, this unun does quite well as a step-down transformer. In matching 50-ohm cable (on the right side) to 25 ohms (A-B) or 22.22 ohms (A-C), the transformation ratios are constant from 1 to 15 MHz. As above, this transformer can easily handle the full legal limit of amateur radio power.

Although the quadrifilar unun shown in the schematic diagram in **Figure 7-2** and in **Photo 7-D** has an impedance transformation ratio of 1.78:1, it should also satisfy many of the 2:1 requirements. This unun, which is designed to match 50-ohm cable to an unbalanced load of 28 ohms, not only has a very broadband response (1 MHz to over 50 MHz), but also offers other possible wideband ratios that will be covered in succeeding chapters.

Specifically, the unun has 5 quadrifilar turns on a

Photo 7-C. Bottom view of the 2:1 unun designed to match 50 ohms to 100 ohms or 112.5 ohms (see Figure 7-1B). The connector is on the low-impedance side.

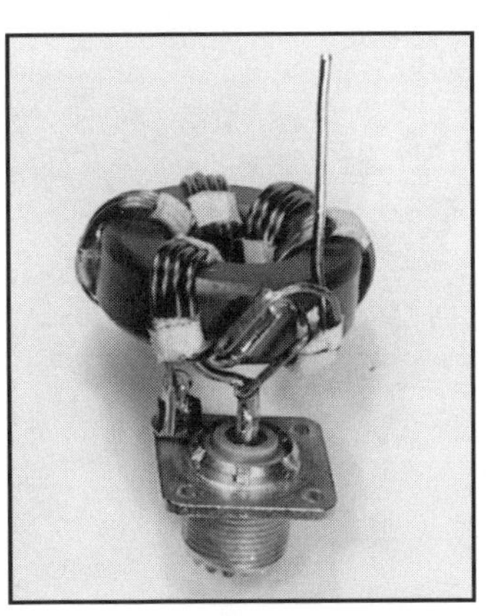

Photo 7-D. Bottom view of the 1.78:1 unun designed to match 50 to 28 ohms. The connector is on the low-impedance side.

1.5-inch OD ferrite toriod with a permeability of 250. Winding 5-6 is No. 14 H Thermaleze wire and the other three are No. 16 H Thermaleze wire. Like the two 2:1 ununs described above, this one also easily handles the full legal limit of amateur radio power.

As with most coiled ununs that have little spacing between adjacent turns, current-crowding (between adjacent turns) can eventually limit the power-handling capability of these devices. It's possible to improve the ability for handling higher currents by using thicker wires, or by using coaxial cables where current-crowding is nonexistent. **Figure 7-3** shows the schematic diagram of a tapped-trifilar transformer that uses two sections of coaxial cable yielding impedance ratios of 2:1 and 2.25:1. **Photo 7-E** shows two trifilar toroidal transformers using low-impedance coaxial cables with their outer braids connected in parallel and acting as the third conductor. These transformers are conservatively rated at 5 kW of continuous power.

The smaller transformer in **Photo 7-E** has 7 trifilar turns of low-impedance coax on a 2-inch OD toroid with a permeability of 290. The No. 14 H Thermaleze wire inner conductors have four layers of Scotch No. 92 tape. The outer braids are made from small coaxial cables (or 1/8-inch tubular braid), and are also wrapped with Scotch No. 92 tape in order to preserve the low characteristic impedances. The inner conductor of the top coax in **Figure 7-3** is tapped at 6 turns from terminal 5. When matching 50 ohms to 22.22 ohms or 25 ohms, the impedance ratio is constant from 1 to over 50 MHz.

The larger transformer in **Photo 7-E** also has 7 trifilar turns, but on a 2.4-inch OD toroid with a permeability of 125. The inner conductors of No. 14 H Thermaleze wire now have a 15-mil wall of Teflon™ sleeving, yielding the low-impedance coaxial cable. The outer braids are the same. Because the ferrite permeability is lower and the lengths of the transmission

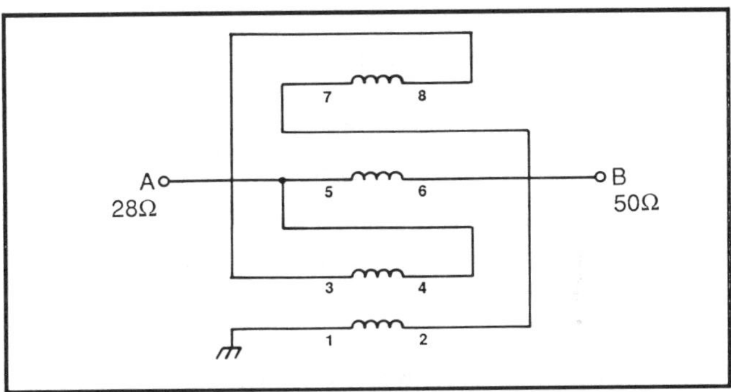

Figure 7-2. Schematic diagram of the quadrifilar unun designed to match 50 to 28 ohms (1.78:1 ratio).

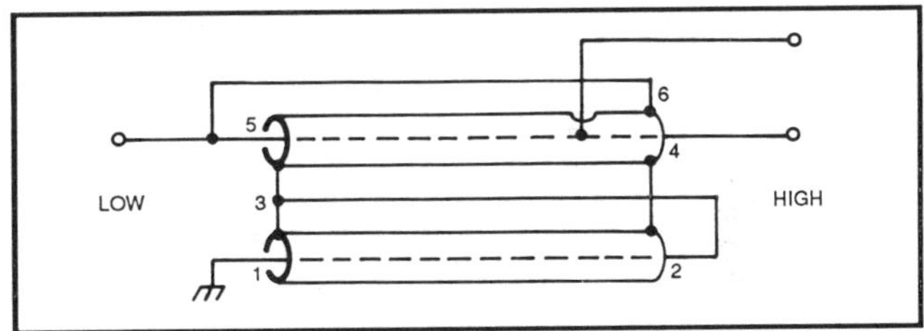

Figure 7-3. Schematic diagram of a tapped-trifilar unun that uses two sections of coaxial cable, yielding impedance ratios of 2:1 and 2.25:1.

lines are longer than those of the smaller unit, this transformer's bandwidth is not quite as good. When matching 50 to 22.22 ohms (this particular unun doesn't have a tapped winding), the impedance ratio is constant from 1.7 to 30 MHz.

Although not shown, these very high-powered 2:1 ununs can also be easily designed to match 50-ohm cable to unbalanced loads of 100 and 112.5 ohms. This is done by using small but high-powered coaxes like RG-303/U, RG-141/U, or RG-142/U.

Sec 7.3 1.5:1 Ununs

Figure 7-4 shows three basic forms of a quintufilar 1.56:1 unun that should satisfy most of the 1.5:1

Photo 7-E. Two toroidal transformers using coaxial cable and designed to match 50 to 22.22 ohms (2.25:1). The smaller transformer is also tapped, yielding a match of 50 to 25 ohms. (2:1).

requirements. As can be seen, the only difference in the schematic diagrams is in the interleaving of the windings. This is done to optimize the performance of these ununs at the various impedance levels. Schematic A is optimized for matching 50 to 75 ohms. Schematic B is optimized for matching 32 to 50 ohms. Schematic C, while optimized for matching 40 to 62 ohms, still yields quite broadband ratios at both 50:75 and 32:50-ohm levels. It should be a useful, general-purpose unun.

Photo 7-F is a photograph of the bottom views (showing the connections) of the three different designs. They appear in the same order as the schematics of **Figure 7-4**; i.e., a) the unun on the left is designed to match 50 to 75 ohms, b) the unun in the center is design to match 32 to 50 ohms, and c) the unun on the right is designed to work quite well at both impedance levels. The SO-239 connectors are all on the low impedance side of the ununs.

All three transformers have four quintufilar turns on a 1.5-inch OD ferrite toroid with a permeability of 250. Their differences are:

1. **50:75 ohms** (on the left on **Figure 7-4** and **Photo 7-F**).

Winding 9-10 is No. 14 H Thermaleze wire. The other four windings are No. 16 H Thermaleze wire. When matching 50 to 75 ohms (actually to 78 ohms), the transformation ratio is constant from 1 to over 30 MHz. In matching 50 ohms (on the right side in **Figure 7-3A**) to 32 ohms, it is still constant from 1 to 15 MHz.

2. **32:50 ohms** (in the center in **Figure 7-4** and **Photo 7-F**).

Winding 5-6 is No. 14 H Thermaleze wire. The other four windings are No. 16 H Thermaleze wire. When matching 32 to 50 ohms, the transformation ratio is constant from 1 to over 30 MHz. In matching 75 ohms (on the right side in **Figure 7-3B**) to 50 ohms, it is still constant from 1 to 15 MHz.

3. **50:75 ohms; 32:50 ohms** (on the right in **Figure 7-4** and **Photo 7-F**).

Winding 7-8 is No. 14 H Thermaleze wire. The other four windings are No. 16 H Thermaleze wire. In matching 32 to 50 ohms, the transformation ratio is constant from 1 to 30 MHz. In matching 75 ohms (on the right side in **Figure 7-3C**) to 50 ohms, it is still constant to 21 MHz. This is quite a good general-purpose design.

Even though a small toriod (with only a 1.5-inch

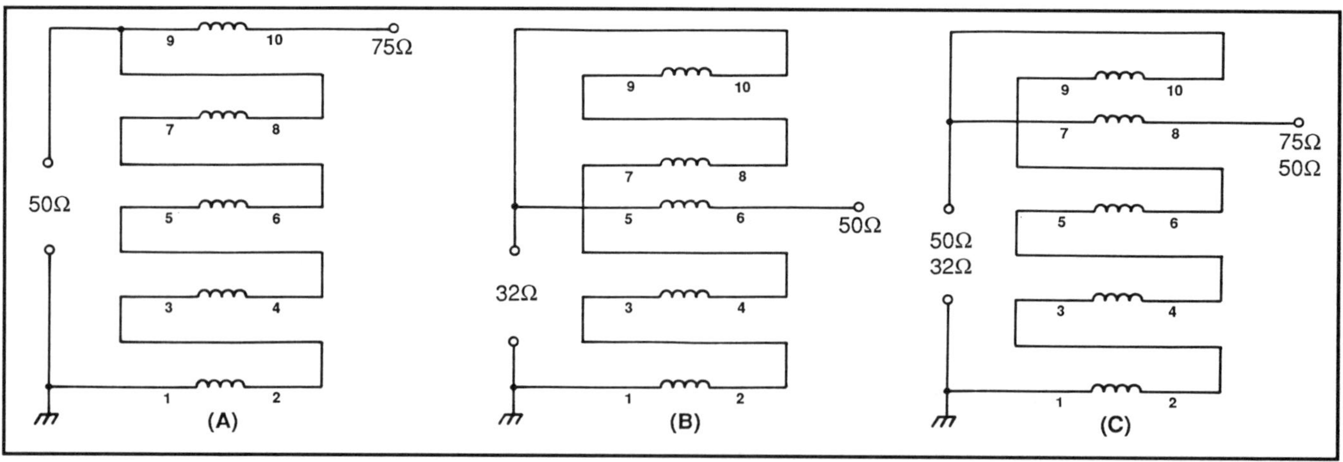

Figure 7-4. Three basic forms of a quintufilar 1.56:1 unun: (A) optimized to match 50 to 75 ohms, (B) optimized to match 32 to 50 ohms, and (C) optimized to match 40 to 62 ohms, resulting in a good general-purpose design.

OD) is used, these ununs are still very sturdy transformers. Because their efficiencies are so high (98 to 99 percent), they can easily handle the full legal limit of amateur radio power.[2] Furthermore, the windings carrying the majority of the current (80 percent) are all No. 14 wire. Only when well-designed ununs are subjected to very high VSWRs, will excessive heating occur. Ununs (and baluns) should never be exposed to these severe conditions.

Sec 7.4 A 1.33:1 Unun

The circuit in **Figure 7-5** evolved after many attempts were made at obtaining a broadband match of 50 to 66.7 ohms (1.33:1). **Photo 7-G** shows the bottom view of an actual design. The SO-239 connector is on the low impedance side. **Photo 7-H** shows the unun mounted in a CU-3015A minibox.

Specifically, this unun has five quintufilar turns on a 1.5-inch OD ferrite toroid with a permeability of 250. Winding 5-6 is No. 14 H Thermaleze wire and is tapped at three turns from terminal 5 (**Figure 7-4**). It is also covered with one layer of Scotch No. 92 polyimide tape, optimizing the performance at the 50:66.7-ohm level. The other four windings are No. 16 H Thermaleze wire.

In matching 50 to 66.7 ohms (A-B), the transformation ratio is practically constant from 1 to 30 MHz. The ratio only decreases by 3 percent across the band. In matching 50 to 32 ohms (C-A), the transformation ratio is constant from 1 to 30 MHz. In matching 75 to 50 ohms (C-A), the ratio is constant from 1 to 15 MHz. In matching 50 to 37.6 ohms (B-A), the ratio is constant from 1 to 15 MHz. As you can see, the unun has some useful broadband multimatches.

As in the cases of the other three ununs, this tapped-unun also easily handles the full legal limit of amateur radio power. Like the 2:1 unun, for higher power capabilities, thicker wires or a three-coax quintufilar design can be used.[2]

Sec 7.5 Construction Tips

Most of my unun designs use the *boot-strap* connection that sums direct voltages (on the high-impedance side) with a delayed voltage, which traverses a single transmission line.[2] Therefore, in order to achieve the very wideband responses, small toroids (which allow the shortest transmission lines) are used. The small 1.5-inch OD toroids offer this advantage. Furthermore, quintufilar (and higher-order windings) ununs are also

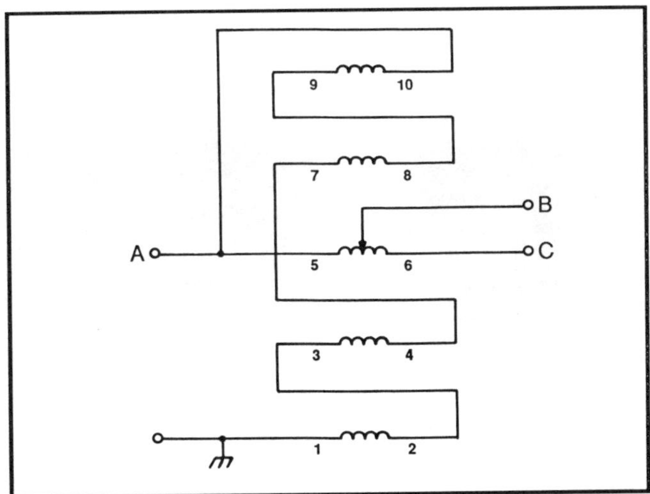

Figure 7-5. Schematic diagram of a quintufilar unun specifically designed to yield a broadband 1.33:1 ratio (66.7:50 ohms; connection B-A). Connection C-A also yields a broadband 1.56:1 ratio (50:32 ohms).

eventually limited in their high-frequency responses by self-resonances. Shorter winding lengths keep these self-resonances well out of the HF band. For the ununs in this chapter, they occur between 45 and 65 MHz, with **Figure 7-4A** having the higher value.

There is also a mechanical advantage in using the smaller toroids. You'll find that the popular CU-3015A minibox makes an excellent enclosure for the 1.5-inch OD toroid. Furthermore, because ferrite is a ceramic and therefore unaffected by moisture, no special precautions need to be taken for out-of-doors use. Potting the transformer in plastic is unnecessary. One must only keep the unun out of a pool of water.

Because well-designed transformers have virtually no flux in the core, their power ratings are mainly determined by the ability of the transmission lines to handle the voltages and currents. Furthermore, it can be shown that the losses in these transformers are related to the voltage gradients along the transmission lines.[2] Thus, they are dielectric-type ferrite losses. This means that the efficiency can be severely degraded with very high VSWRs since higher voltage gradients occur under these conditions.

Several suggestions can be made regarding the construction of ununs using these higher-order windings (trifilar, quadrifilar, etc.). They are:

1) Make a ribbon out of the wires and wind them all at the same time. This keeps the wires as close as possible, resulting in the maintenance of the optimum characteristic impedance of the transmission lines. I found that strips made with 1.25 by 0.375 to 0.55 inch

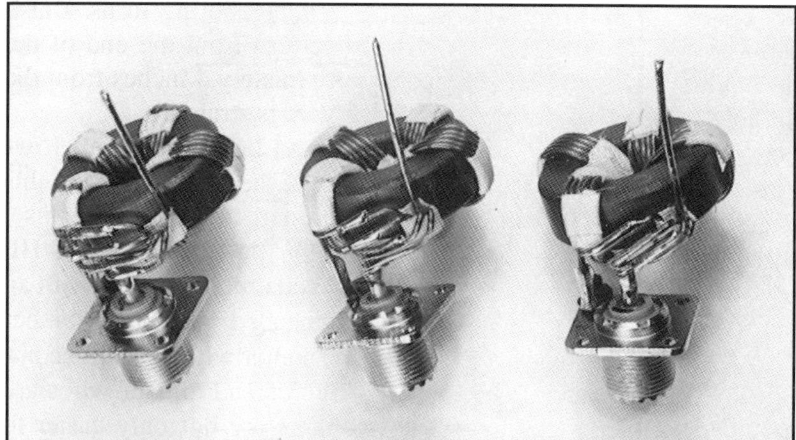

Photo 7-F. Photograph of the three different unun designs shown in Figure 7-4: A) on the left, 50:75 ohms, B) in the center 32:50 ohms, and C) on the right, a general purpose design matching both impedance levels quite well.

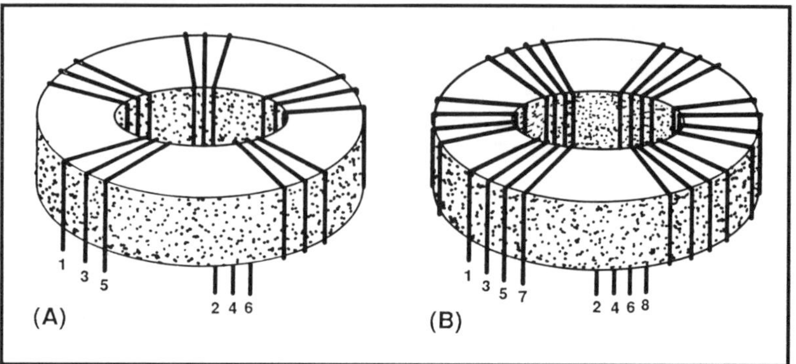

Figure 7-6. Pictorials of higher-order windings: (A) trifilar and (B) quadrifilar.

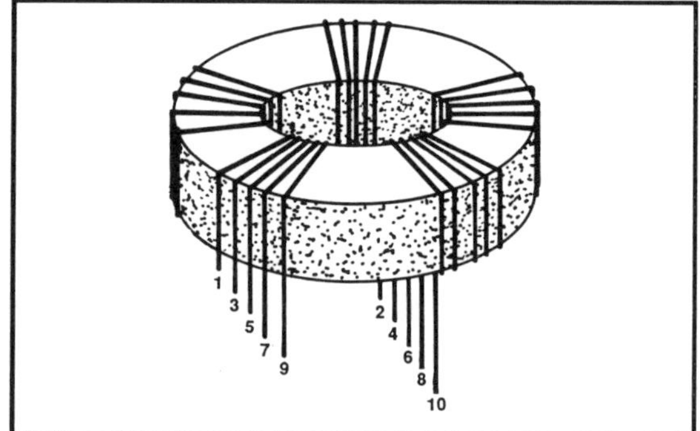

Figure 7-7. Pictorial of a quintufilar winding.

glass tape (Scotch No. 27), clamped about every 1/2 inch, hold the wires in place very well. The starting lengths of the wires should be about 5 inches longer than one would calculate knowing the number of turns and the length around each turn.

2) Because work-hardening of the copper wire takes place in coiling it around a toroid, a pair of pliers and a strong thumb (and arms) are indispensable tools. It takes considerable effort to wind these transformers. Also, because these designs have adequate margins at their low-frequency ends, some space between the windings and the toroids can be tolerated.

3) It is helpful to recognize the various patterns that appear at the ends of the windings. **Figure 7-6** shows a drawing of the trifilar and quadrifilar patterns and **Figure 7-7** shows the quintifilar pattern. Note that terminal 1 and terminal 6 or terminal 8 or 10 are the outside terminals of the patterns. Also, note that terminal 1 is always grounded in the schematic diagrams.

4)Tapping windings can be one of the more difficult tasks in constructing these transformers. Winding a tapped transformer is also more difficult. I found that the edge of a small, fine file does the best job in removing the insulation. About 1/8 to 1/4 inch is removed around the wire. It also helps to remove some of the copper. Then a flat 1/8-inch copper strip or No. 14 wire flattened on one end is soldered to the bare wire. The soldered connection is then rendered smooth using the edge of the file. Finally, two pieces of Scotch No. 92 polyimide tape are placed on the

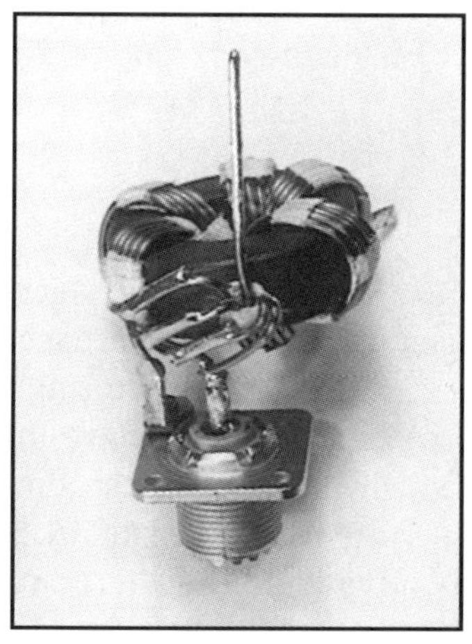

Photo 7-G. Photograph of the 1.33:1 unun (66.7:50 ohms). This transformer also has a broadband 1.56:1 ratio (50:32 ohms).

Photo 7-H. The 1.33:1 unun mounted in a CU-3015A minibox.

joint to insulate it from the neighboring turns. I also found that a tap placed one-turn from the end of the winding is best made approximately 4 inches from the end of the wire (when the wire is straight).

Finally, a comment should be made about low-power ununs. Practically all of the transformers in this chapter can be easily designed for low-power use. Designs capable of handing the full output of HF transceivers can be readily constructed. Cores with an OD of 1.25 inch are recommended. The same number of turns, but with one size smaller wire, is also recommended. Because smaller cores and thinner wires are used, these lower-power units are not only easier to construct, but they also have wider bandwidths due to the shorter lengths of the windings.

Chapter 8

Dual-Ratio Ununs

Sec 8.1 Introduction

Earlier chapters in this book on ununs have basically covered single-ratio transformers. Broadband ratios of 1.33:1, 1.5:1, 2:1, and 4:1 were the design objectives. The 1.33:1 ratio was obtained by tapping a 1.5:1 (actually 1.56:1) quintufilar-wound unun. The 2:1 ratio was obtained by tapping a 2.25:1 trifilar-wound unun. Although these transformers can be considered to have two broadband ratios (1.33:1 and 1.56:1 or 2:1 and 2.25:1), their two ratios were not different enough for many practical applications. This is especially true of antennas where the input impedance varies with frequency.

My earlier work,[2] and a recent article by Genaille,[10] have shown that a host of ratios (less than 4:1) can be obtained by tapping the bifilar winding of a Ruthroff 4:1 unun.[9] However, the bandwidths obtained using this technique are quite limited with each ratio and are highly dependent upon the impedance level. This is particularly true when a rod core is used because it requires more turns (resulting in longer transmission lines) in order to obtain the necessary choking reactance that isolates the input from the output. Furthermore, ratios around 2:1 exhibit more loss because autotransformer action enters into the matching process.[2]

By using quadrifilar and quintufilar windings on small 1.5-inch OD cores, and connecting them in such a way that the characteristic impedances of the windings are near optimum, two very different and broadband ratios matching 50 ohms to lower impedances are obtained. Furthermore, because the transmission lines in these transformers are so very short (8 to 9 inches in length), these transformers do quite well in matching 50 ohms to higher impedances (as step-up transformers).

This chapter presents two ununs which have two

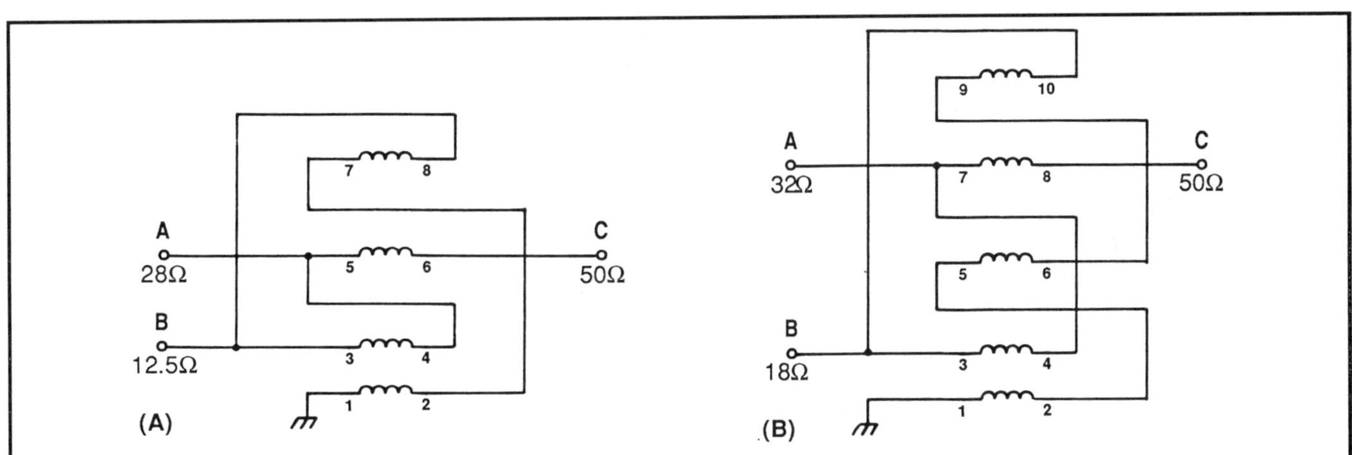

Figure 8-1. Schematic diagrams of dual-ratio ununs: (A) 1.78:1 connection C-A, 4:1 connection C-B; (B) 1.56:1 connection C-A, 2.78:1 connection C-B.

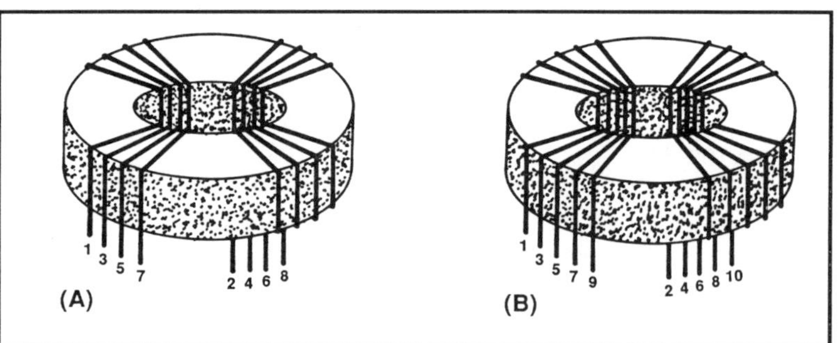

Figure 8-2. Pictorials of higher-order windings: (A) quadrifilar, (B) quintufilar.

broadband ratios that differ by a factor of two! One has a 1.5:1 and a 3:1 ratio (actually 1.56:1 and 2.78:1). The other has a 2:1 (actually 1.78:1) and a 4:1 ratio. Also, this chapter introduces the novel technique of connecting these two transformers in parallel on their 50-ohm sides, resulting in four very broadband ratios.

Sec 8.2 2:1 and 4:1 Ratios

Figure 8-1A shows the schematic diagram of the quadrifilar unun yielding ratios of 2:1 (actually 1.78:1) and 4:1. **Figure 8-2A** provides a pictorial of its windings. On the left in **Photo 8-A**, you see the bottom view of an unun. The cable connector is on the 28-ohm (1.78:1) side. This unun has four quadrifilar turns of No. 14 H Thermaleze wire on a 1.5-inch OD toroid with a permeability of 250.

In matching 50 to 28 ohms (connection C-A), the response is flat to within 1 percent from 1 to 30 MHz. From 1 to 50 MHz, it's flat to within 2 percent. In matching 50 to 12.5 ohms (connection C-B), the response is flat to within 3 percent from 1 to 30 MHz.

Because very short transmission lines are used, this transformer performs quite well as a step-up transformer. In matching 50 to 200 ohms (connection B-C), the response is flat to within 3 percent from 1.5 to 10 MHz. In matching 50 to 89 ohms (connection A-C), the response is flat to within 5 percent from 1.5 to 30 MHz!

Sec 8.2.1 Construction Tips

Start with about 14 inches of straightened wire. Form the wires into a ribbon with clamps of Scotch No. 27 glass tape every 1/2 inch. I found that strips 3/16 inch wide and about 1.5 inches long do a good job. The clamps should be long enough to go around the wires twice. After winding, connect terminals 2 and 7. Then connect terminals 3 and 8. Finally, connect terminals 4 and 5. Because work-hardening takes place quickly, you will find that a pair of pliers and a strong thumb

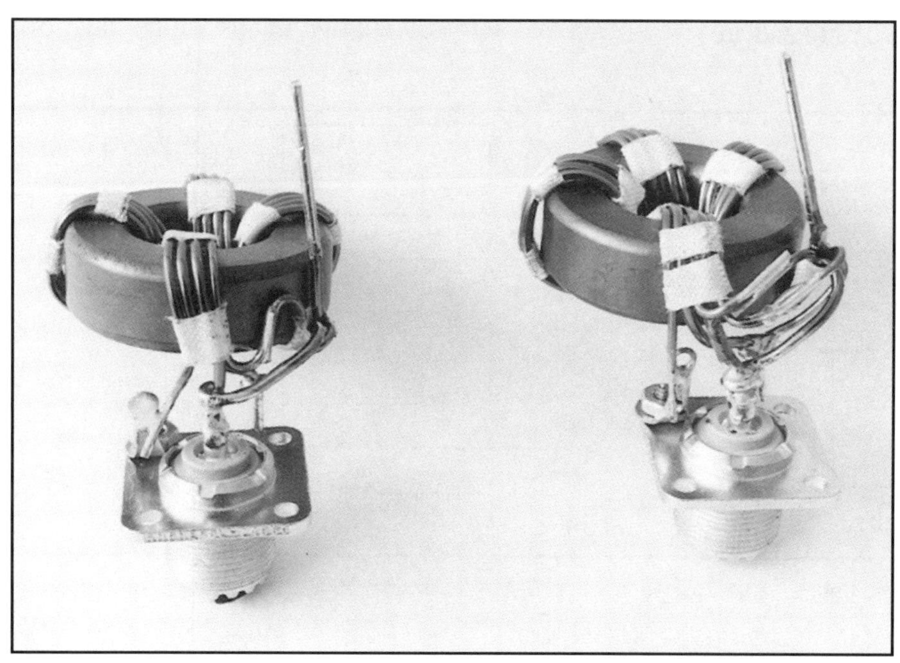

Photo 8-A. On the left, a quadrifilar unun with ratios of 1.78:1 and 4:1; on the right, a quintufilar unun with ratios of 1.56:1 and 2.78:1.

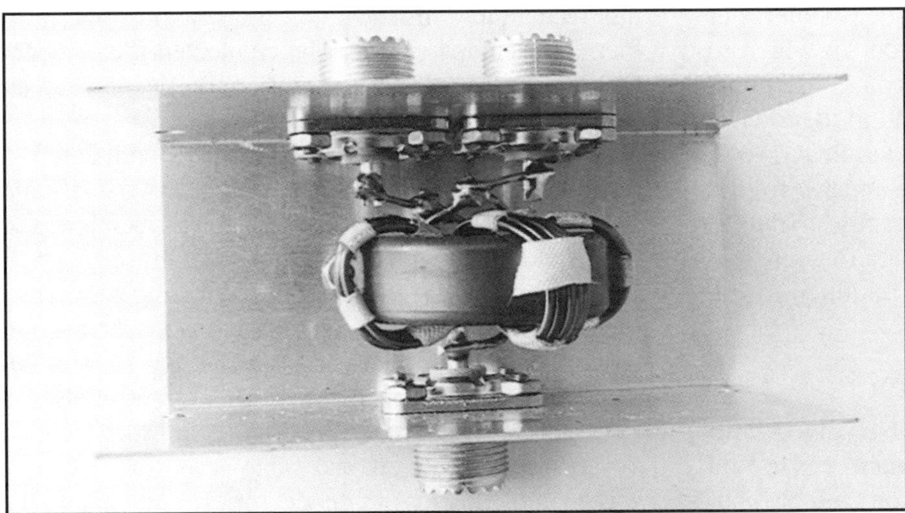

Photo 8-B. The dual-ratio quintufilar unun mounted in a 4-inch long by 2-inch wide by 2.75-inch high minibox.

(and arms) are necessary tools. You will also find that winding these exceptional performing transformers is not easy. As in all endeavors, practice really pays off.

Sec 8.3 1.5:1 and 3:1 Ratios

Figure 8-1B shows the schematic diagram of the quintufilar unun yielding ratios of 1.5:1 and 3:1 (actually 1.56:1 and 2.78:1). **Figure 8-2B** is a pictorial of its windings. On the right side in **Photo 8-A** you see a bottom view of an unun. The cable connector is on the 32-ohm (1.56:1) side. This unun has four quintufilar turns on a 1.5-inch OD toroid with a permeability of 250. Windings 3-4 and 7-8 are No. 14 H Thermaleze wire. The other three are No. 16 H Thermaleze wire. Winding 7-8 also has two layers of Scotch No. 92 polyimide tape, which optimizes the 1.56:1 ratio.

In matching 50 to 32 ohms (connection C-A), the response is essentially flat (less than 1 percent variation) from 1 MHz to over 40 MHz. Without the two layers of Scotch No. 92 tape, the response varies by 4 percent from 1 to 30 MHz. When used as a step-up transformer matching 50 to 78 ohms (connection A-C), and with the two layers of Scotch No. 92 tape on winding 7-8, the response is flat to within 5 percent from 1 to 15 MHz. Without the extra insulation on winding 7-8, the response is flat to within 5 percent from 1 to 7.5 MHz.

In matching 50 to 18 ohms (connection C-B), the variation in response is less than 3 percent from 1 to 40 MHz. The response is the same whether winding 7-8 is covered with the extra insulation or not. As a step-up transformer matching 50 to 139 ohms (connection B-C), the response is flat to within 3 percent from 1 to 10 MHz (with or without the extra insulation on winding 7-8). **Photo 8-B** shows this unun mounted in a 4 inch long by 2 inch wide by 2.75 inch high minibox. The two cable connectors on the low impedance side could be replaced with feedthrough insulators for antenna use.

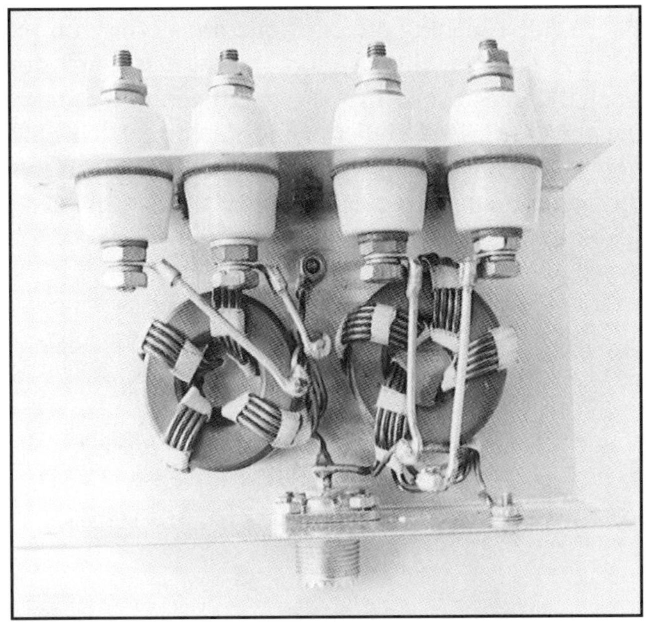

Photo 8-C. The two dual-ratio ununs connected in parallel on their 50-ohm sides providing four broadband ratios close to 1.5:1, 2:1, 3:1, and 4:1. The quadrifilar unit is on the left and the quintufilar unit is on the right. The enclosure is a 5 inch long by 3 inch wide by 2 inch high minibox.

Sec 8.3.1 Construction Tips

Prepare the ribbon as was described for the quadrifilar unun. If you choose to use the two extra layers of Scotch No. 92 tape on winding 7-8, make sure this winding is on the outside position of the ribbon (refer to **Figure 8-2B**). I found the best order in which to connect the wires is as follows: first, connect terminal 2 to 5; second, connect terminal 6 to 9; third, connect terminal 3 to 10; and finally, connect terminal 4 to 7. As was mentioned before, because a small toroid is used in order to achieve the best response (due to shorter transmission lines), this transformer also requires considerable strength and patience in the winding process.

Sec 8.4 Parallel Transformers

One of the most pleasant surprises I received with these efficient and broadband transformers was to find that they can be connected in parallel on their 50-ohm sides and still possess the same performance levels. Because the loading effect of one transformer on the other is minimal (like a short length of transmission line), the transformer that is properly terminated takes the power, while the other one is transparent. This lets you obtain four very wideband ratios with the two dual-ratio ununs described in this chapter. Obviously, this technique eliminates one transmission line. Furthermore, baluns can also be connected this way for feeding, with a single coaxial cable, beams and dipoles with different resonant impedances.

Photo 8-C shows two transformers connected in parallel on their 50-ohm sides using the two dual-ratio ununs described in this chapter. As was mentioned, it now yields four wideband ratios very close to 1.5:1, 2:1, 3:1, and 4:1. I have used this matching network to feed a host of ground-fed antennas (over a good ground system). In one case, I had a 10, 15, and 20-meter trap vertical, slopers for 40 and 160 meters, a 12-meter vertical, and an inverted L for 80 meters all matched to a single coaxial cable at the same time. It was a simple matter of connecting each antenna to the output terminal that presented the best match (lowest VSWR). This technique is actually an extension of connecting dipoles for different bands, in parallel. The antenna that presents the correct impedance takes the power, and the others are essentially transparent. In many cases, I found that only one transformer with two broadband ratios performed adequately.

Chapter 9

Multimatch Ununs

Sec 9.1 Introduction

Broadband multimatch ununs capable of high-power applications have been the goal of many designers over the years. Some have resorted to using conventional autotransformers with tapped windings to obtain the many impedance transformation ratios. However, these attempts met with little success because of the device's limited bandwidths and efficiencies. Others (including myself)[2] have tried tapping a bifilar Ruthroff unun.[9] Although these designs yielded the high efficiencies of transmission line transformers, they had limited bandwidths. Furthermore, their best bandwidths (for the various ratios) occurred at odd impedance levels. In other words, they didn't meet the objective of broadband operation with one of the input or output ports at 50 ohms.

Chapter 8 presented two ununs which had two broadband ratios that differed by a factor of two. One had a 1.5:1 and a 3:1 ratio (actually 1.56:1 and 2.78:1). The other had a 2:1 (actually 1.78:1) and a 4:1 ratio. This chapter describes two multimatch designs that are capable of many more broadband ratios. For the most part, both are capable of broadband operation from 1.7 to 30 MHz.

One unun has the following five ratios (which are close to): 1.5:1, 2:1, 4:1, 6:1, and 9:1. Because the two lower ratios work well in either direction (that is stepping up or down from 50 ohms), this design can match 50-ohm cable to impedances as high as 100 ohms (actually 112.5 ohms) and as low as 5.6 ohms over the frequency range. As a result, it has seven usable applications. Furthermore, because this is a transmission line transformer that cancels out the flux in the core, losses (in a matched condition) of only 0.04 to 0.08 dB can be expected.

The novelty in this design lies in the use of a trifilar winding (with one winding tapped) on a very small ferrite toroid, resulting in the shortest possible lengths of transmission lines. The windings are also connected in such a manner as to optimize their characteristic impedances from an overall standpoint.

I have used the adjective *ultimate* to describe the second unun design. Although it might be risky business, I assume that this design will meet one of the most common definitions for this adjective—namely, *beyond which it is impossible to go*. For many of us, the classic use of this adjective was made by Lew McCoy in describing his popular transmatch.[22] Al-

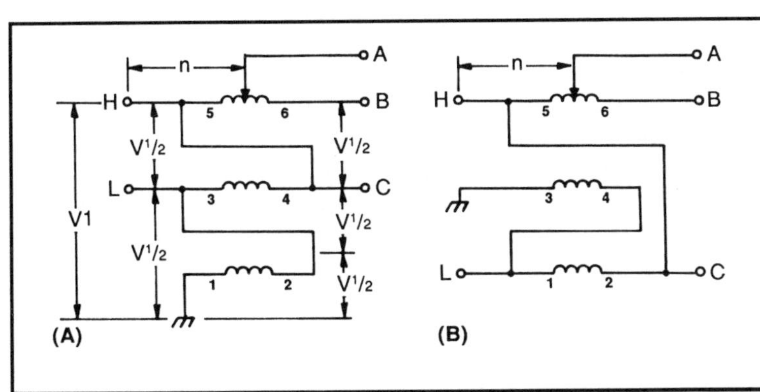

Figure 9-1. Circuit diagrams for the 5-ratio unun: (A) diagram for analysis; (B) transposed windings for best overall performance.

Photo 9-A. Bottom view of the 5-ratio unun of Figure 9-1B. The upper-left lead is terminal C. The upper-right lead is terminal B. The lower-left lead is terminal H. The lead pointing straight down is grounded (terminal 3). The lower-right lead is terminal L.

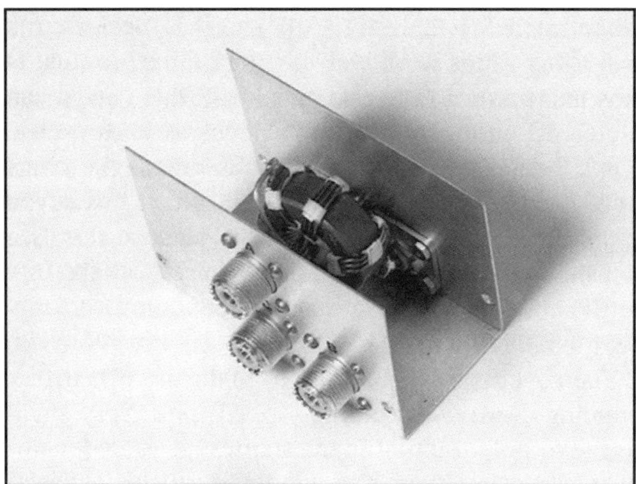

Photo 9-B. The high-power unit mounted in a 4 inch long by 2 inch wide by 2.75 inch high minibox.

though there have been some improvements to Lew's design, his use of this definite (and strong) adjective can be said to have withstood the test of time. I hope my use meets with similar success.

While the tapped-trifilar design provides five broadband ratios and seven practical applications, the *ultimate* design presented in this chapter goes well beyond this number. It uses a tapped-quadrifilar design that yields the following ten broadband ratios: 1.36:1, 1.56:1, 1.78:1, 2.25:1, 3.06:1, 4:1, 6.25:1, 9:1, 12.25:1, and 16:1. Because the four lower ratios also work quite well in either direction, this design offers *fourteen* applications in matching 50-ohm cable to impedances as high as 112.5 ohms and as low as 3.125 ohms. It also has the advantage of using a small low-loss toroidal core. Additionally, the windings are also interleaved in a pattern that optimizes their characteristic impedance.

This achievement comes at a price called *difficulty,* however. The 5-ratio unun, which uses a trifilar winding, is considerably easier to wind. In addition, the quadrifilar 10-ratio unun has two of its windings tapped, while the 5-ratio unun has only one (see **Chapter 7** on tapping windings). If you have had little experience in winding ununs or baluns, attempt simplified versions of these two multimatch ununs first. These versions eliminate the tapping of the windings. For the trifilar unun, the remaining ratios would be: 2.25:1, 4:1, and 9:1. For the quadrifilar unun, they would be: 1.78:1, 2.25:1, 4:1, 9:1, and 16:1.

For those interested in the design considerations of these broadband multimatch transformers, a brief review is presented in each section. These sections are followed by others describing high-power designs capable of handling the full legal limit of amateur radio power. Finally, the remaining sections present low-power designs capable of handling the output of any HF transceiver. Because transmission line transformers can be made so efficient in matching 50 to 100 ohms or less, their small sizes will surprise many readers. Therefore, the combination of using small ferrite toroids with the maximum allowable permeability (less than 300) for high efficiency,[2] and with sufficient turns to meet the low-frequency objective, results in the excellent performance exhibited by the designs in this chapter.

Sec 9.2 The 5-Ratio Unun

Let's first look at **Figure 9-1A** because it is the easiest form of the trifilar-wound unun to explain. For example, if the input voltage to ground, V_1, is connected to terminal H, the output terminal B, has a voltage to ground of $3/2V_1$. This results in a transformation ratio, g, of $(3/2)^2$ or 2.25:1. This should satisfy most 2:1 requirements. If the output is at terminal A to ground, then the output voltage is:

$$\begin{aligned} V_o &= V_1 + V_1(n/2N) \\ &= V_1(1 + n/2N) \end{aligned} \quad \text{(Eq 9-1)}$$

where:

N = the total number of turns on the winding

n = the number of turns from terminal 5.

The transformation ratio, g, then becomes:

$$\begin{aligned} g &= (V_o/V_1)^2 \\ &= (1 + n/2N)^2 \end{aligned} \quad \text{(Eq 9-2)}$$

If the input voltage to ground, V_1, is connected to terminal L, then terminal C has twice the voltage of V_1—resulting in a 4:1 ratio. Terminal B has three times the voltage resulting in a 9:1 ratio. With terminal A, the output voltage is:

$$V_o = 2V_1 + V_1(n/N)$$
$$= V_1(2 + n/N) \quad \text{(Eq 9-3)}$$

The transformation ratio, g, then becomes:

$$g = (2 + n/N)^2 \quad \text{(Eq 9-4)}$$

Sec 9.2 A High-power 5-Ratio Unun

After several attempts at rearranging the windings of **Figure 9-1A** for the best overall performance (optimizing the effective characteristic impedances of the windings), **Figure 9-1B** evolved. **Photo 9-A** shows the bottom view of an unun, using the circuit of **Figure 9-1B**, capable of handling the full legal limit of amateur radio power. **Photo 9-B** shows the unit mounted in a CU-3015A minibox. It has five trifilar turns on a 1.5-inch OD ferrite toroid with a permeability of 250. Winding 5-6 is tapped at two turns (n = 2) from terminal 5.

If the 9:1 ratio matching 50 to 5.6 ohms (connection B-L) is to be used at full power, then winding 3-4 should be No. 12 H Thermaleze wire. If not, then all windings can be No. 14 H Thermaleze wire.

A listing of the expected performance across the band from 1.7 MHz to 30 MHz, with the various ratios, is as follows:

9:1 (B-L); 50:5.6 ohms
Ratio is within 1 percent!
5.75:1 (A-L); 50:8.7 ohms
Ratio decreases by 5 percent.
4:1 (C-L); 50:12.5 ohms
Ratio increases by 15 percent (the greatest deviation of all the ratios).
2.25:1
a) (B-H); 50:22.22 ohms
Ratio decreases by 4 percent.
b) (H-B); 50:112.5 ohms
Ratio increases by 8 percent.
1.44:1
a) (A-H); 50:35 ohms
Ratio decreases by 10 percent.
b) (H-A); 50:72 ohms
Ratio increases by 2 percent.

Several comments should be made regarding the expected results shown above. First of all, the greatest deviation from a flat response at any ratio occurs when matching 50 to 12.5 ohms (connection C-L; a 4:1 ratio). If an accurate insertion loss measurement was made at this ratio and impedance level, the result would show an insignificant difference across the band. Secondly, the major part of the deviations for all ratios occurs beyond 15 MHz (the effect of standing waves). Finally, the higher ratios should *never* be used to match 50 ohms to 450 ohms, 288 ohms, and 200 ohms, respectively. The characteristic impedances and choking reactances *do not* allow for broadband operation under these conditions.

Photo 9-C. The low-power unit mounted in a homemade 2-inch long by 1.5-inch wide by 2.25-inch high minibox.

Sec 9.2.2 A Low-power 5-Ratio Unun

Photo 9-C shows a low-power unit mounted in a homemade 2 inch long by 1.5 inch wide by 2.25 inch high minibox. It has six trifilar turns of No. 16 H Thermaleze wire on a 1.25-inch OD ferrite toroid with

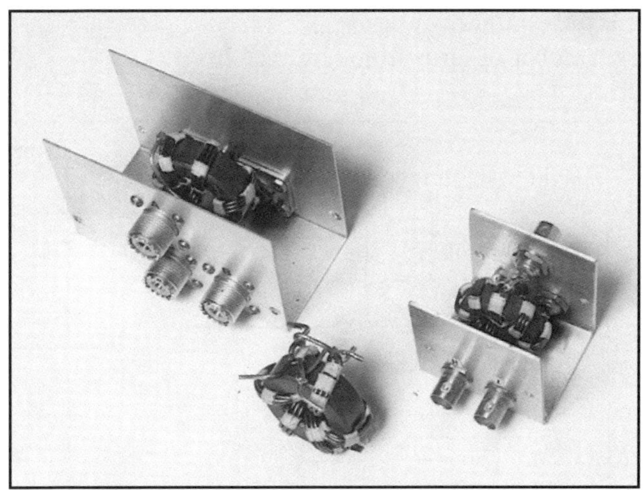

Photo 9-D. The three 5-ratio ununs together. From left to right, the high-power unit mounted and unmounted, the low-power unit.

a permeability of 250. The tap on winding 5-6 is located three turns from terminal 5, yielding ratios of 6.25:1 and 1.56:1 instead of the 5.75:1 and 1.44:1 ratios of the high-power unit. In actual use, these differences should be negligible.

Because this unun has shorter transmission lines than its high-power counterpart, the deviations of the ratios across the band are even smaller. It is also interesting to note that, if No. 14 H Thermaleze wire was used in winding 3-4, this very small unun could very well be rated at 500 watts of continuous power!

Photo 9-D shows all three 5-ratio ununs together.

Sec 9.3 The 10-Ratio Unun— The Ultimate Multimatch

Figure 9-2A is presented here because it is the easiest form of a quadrifilar-wound unun to explain. With the input voltage, V_1, connected to the various terminals on the left (the low-impedance side), and with very short transmission lines compared to the wavelength, we have the following transformation ratios:

1. V_1 connected to terminal A

a) At terminal D the output voltage V_o is $4/3V_1$. Therefore, the transformation ratio, g, with connection A-D is:

$$g = (4/3)^2 = 1{:}1.78 \qquad \text{(Eq 9-5)}$$

b) At terminal F the output voltage is:

$$V_o = V_1(1 + n/3N) \qquad \text{(Eq 9-6)}$$

where:
N = total number of turns
n = number of turns from terminal 7.

The transformation ratio with connection A-F then becomes:

$$g = (V_o/V_1)^2 = (1 + n/3N)^2 \qquad \text{(Eq 9-7)}$$

2. V_1 connected to terminal B

a) At terminal E the output voltage is $3/2V_1$. Thus, the transformation ratio with connection B-E is:

$$g = (3/2)^2 = 1{:}2.25 \qquad \text{(Eq 9-8)}$$

b) At terminal G the output voltage is:

$$V_o = V_1(1 + n/2N) \qquad \text{(Eq 9-9)}$$

where n = number of turns from terminal 5.

The transformation ratio with connection B-G becomes:

$$g = (V_o/V_1)^2 = (1 + n/2N)^2 \qquad \text{(Eq 9-10)}$$

c) At terminal D the output voltage is $2V_1$. The transformation ratio with connection B-D is:

$$g = (2)^2 = 1{:}4 \qquad \text{(Eq 9-11)}$$

d) At terminal F the output voltage is:

$$V_o = V_1(3/2 + n/2N) \qquad \text{(Eq 9-12)}$$

where n = number of turns from terminal 7.

The transformation ratio with connection B-F, then, is:

$$g = (V_o/V_1)^2 = (3/2 + n/2N) \qquad \text{(Eq 9-13)}$$

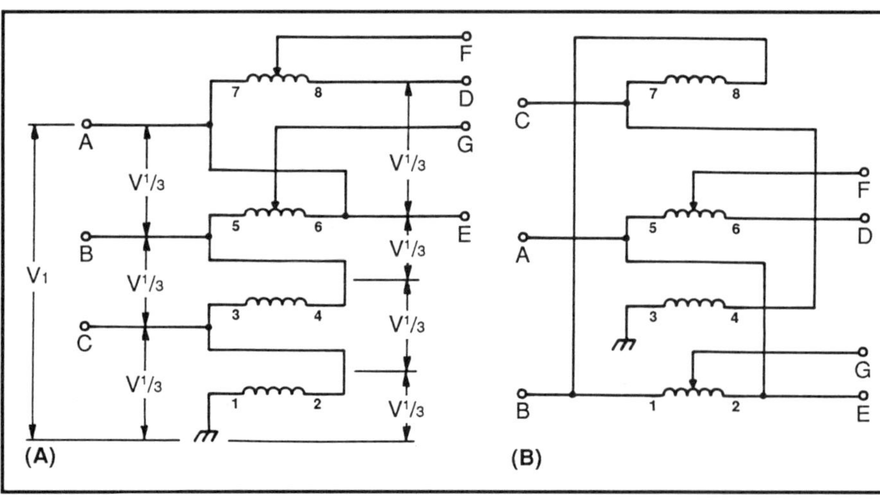

Figure 9-2. Circuit diagrams for the 10-ratio unun: (A) diagram for analysis; (B) transposed windings for best overall performance.

3. V_1 connected to terminal C

a) At terminal E, the output voltage is $3V_1$. The transformation ratio with connection C-E becomes:

$$g = (3)^2 = 1{:}9 \qquad \text{(Eq 9-14)}$$

b) At terminal G, the output voltage is:

$$V_o = V_1(2 + n/N) \qquad \text{(Eq 9-15)}$$

where n = number of turns from terminal 5.

The transformation ratio with connection C-G is:

$$g = (V_o/V_1)^2 = (2 + n/N)^2 \qquad \text{(Eq 9-16)}$$

c) At terminal D, the output voltage is $4V_1$. The transformation with connection C-D becomes:

$$g = (4)^2 = 1{:}16 \qquad \text{(Eq 9-17)}$$

d) At terminal F the output voltage is:

$$V_o = V_1(3 + n/N) \qquad \text{(Eq 9-18)}$$

where n = number of turns from terminal 7.

The transformation ratio with connection C-F is:

$$g = (V_o/V_1)^2 = (3 + n/N)^2 \qquad \text{(Eq 9-19)}$$

Sec 9.3.1 A High-power 10-Ratio Unun

Figure 9-2B evolved after several attempts at rearranging the windings of **Figure 9-2A** for best overall performance (optimizing the effective characteristic impedances of the windings). **Photo 9-E** shows the bottom view of an unmounted unun using the circuit of **Figure 9-2B**. The top-left lead is terminal E. The top-right lead is terminal D. The bottom-left lead is terminal B. The center lead (connected to the SO-239 connector) is terminal A.

The bottom-right lead is terminal C. Below these three bottom leads is a ground connection (terminal 3 in **Figure 9-2B** to the SO-239 connector. **Photo 9-F** shows three different views of this high-power unit mounted in a 4 inch long by 2 inch wide by 2.75 inch high CU-3015A minibox.

This 10-ratio unun has four quadrifilar turns of No. 14 H Thermaleze wire on a 1.5-inch OD ferrite toroid with a permeability of 250. Winding 5-6 is tapped at 2

Photo 9-E. Bottom view of the 10-ratio unun. The connector is on terminal A.

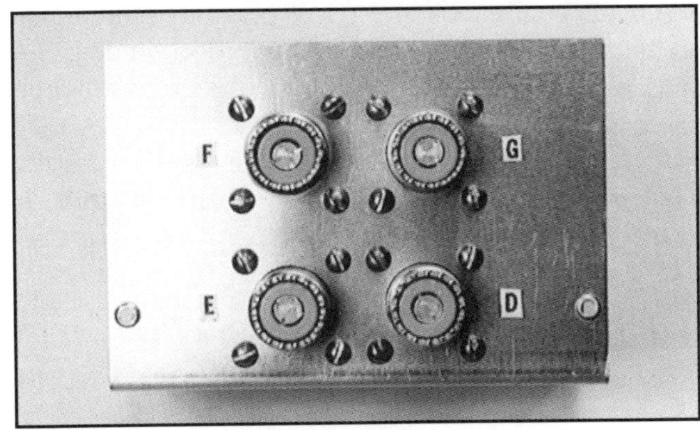

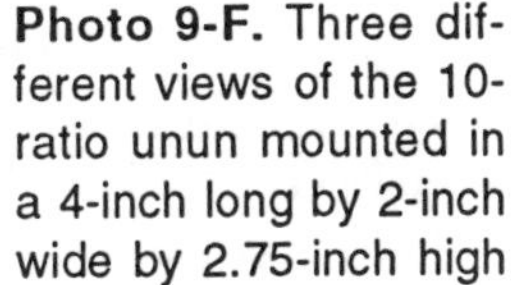

Photo 9-F. Three different views of the 10-ratio unun mounted in a 4-inch long by 2-inch wide by 2.75-inch high CU-3015A minibox.

Photo 9-G. The low-power unit mounted in a homemade 2.25-inch long by 1.5-inch wide by 2.25-inch high minibox.

turns from terminal 5 and winding 7-8 is tapped at 2 turns from terminal 7.

If the 9:1 ratio (connection C-E), the 12.25:1 ratio (connection C-F), and the 16:1 ratio (connection C-D), are to be used at the full legal limit of amateur radio power, then I suggest that winding 3-4 be replaced with No. 12 H Thermaleze wire. If not, then these three ratios should be used at lower power levels (of 500 watts continuous and 1 kW peak). It should also be mentioned that using No. 12 wire for winding 3-4 adds a greater degree of *difficulty* to the construction process.

A listing of the expected performance across the band from 1.7 to 30 MHz, with the various ratios, is as follows:

16:1 (D-C); 50:3.125 ohms
Ratio is constant up to 21 MHz. It then decreases by 15 percent.
12.25:1 (F-C); 50:4.08 ohms
Ratio is constant.
9:1 (E-C); 50:5.56 ohms
Ratio increases by 5 percent.
6.25:1 (G-C); 50:8 ohms
Ratio is constant.
4:1 (D-B); 50:12.5 ohms
Ratio decreases by 5 percent.
3.06:1 (F-B); 50:16.3 ohms
Ratio decreases by 10 percent.
2.25:1
a) (E-B); 50:22.22 ohms
Ratio increases by 4 percent.
b) (B-E); 50:112.5 ohms
Ratio increases by 50 percent (the greatest deviation across the band of any of the ratios).
1.78:1
a) (D-A); 50:28.1 ohms.
Ratio is constant.
b) (A-D); 50:89 ohms.
Ratio increases by 15 percent.
1.56:1
a) (G-B); 50:32 ohms
Ratio increases by 10 percent.
b) (B-G); 50:78 ohms
Ratio increases by 40 percent.
1.36:1
a) (F-A); 50:36.8 ohms
Ratio decreases by 9 percent.
b) (A-F); 50:68 ohms
Ratio increases by 1.5 percent.

Sec 9.3.2 A Low-power 10-Ratio Unun

Photo 9-G shows a low-power unit mounted in a homemade 2.25 inch long by 1.5 inch wide by 2.25 inch high minibox. It has five quadrifilar turns of No. 16 H Thermaleze wire on a 1.25-inch OD ferrite toroid with a permeability of 250. The tap on winding 5-6 (**Figure 9-2B**) is at three turns from terminal 5 and on winding 1-2; it is three turns from terminal 1. Because the number of turns are different from the high-power unit, so are the ratios that use the taps. In this case, they are little larger. Specifically, the tapped ratios are now: 1:12.96, 1:6.76, 1:3.24, 1:1.69, and 1:1.44. If the taps were at two turns from terminals 5 and 1, the ratios would be a little less than those of the high-power unit. You can play with the equations in the first section of this chapter and arrive at many different ratios.

Because this unun has shorter transmission lines than its high-power counterpart, the deviations of the ratios across the HF band are generally smaller. Also, if winding 3-4 (in **Figure 9-2B**) were replaced with No. 14 H Thermaleze wire, this low-power unit could very well be rated at 500 watts of continuous power for all ratios!

Chapter 10

Ununs for Beverage Antennas

Section 10.1 Introduction

The Beverage antenna[19] is well known by 160-meter enthusiasts for enhanced signal-to-noise ratios when there are high levels of interference and atmospheric noise. If erected properly, Beverages also have excellent directivity. However, they are quite inefficient and therefore not generally suitable as transmitting antennas. Important considerations with Beverages are the terminating resistor (for the more common single-wire version) and the input matching unun (unbalanced-to-unbalanced transformer). The terminating resistor and the impedance ratio of the unun are determined by the characteristic impedance of the antenna acting as a long transmission line with one good conductor and one poor conductor (the earth). This line is generally between 400 and 600 ohms, and theoretically given by:

$$Z_0 = 138 \times \log(4h/d) \qquad \text{(Eq 10-1)}$$

where:

Z_0 = characteristic impedance of the Beverage

h = height of the wire above ground

d = diameter of the wire.

This chapter presents low- and high-power versions of multimatch ununs designed to match 50-ohm cable to unbalanced loads from 450 to 800 ohms. The low-power unit, which is capable of handling continuous power levels up to 100 watts, is specifically designed for the Beverage antenna when it is performing as a receiving antenna. The high-power unit, which is capable of handling 1 kW of continuous power, can be used with the Beverage or any other traveling wave antenna when used as a transmitting antenna. Also presented are high-power designs capable of flat response, including the entire AM broadcast band. These multimatch ununs could be of interest to designers of high-power amplifiers for the broadcast band. A little theory on how these devices are designed is also provided.

Sec 10.2 A Little Theory

Transmission line transformers[2] (the unun being a subset thereof) are known for having greater bandwidths and efficiencies than their counterparts, the conventional transformers. Design considerations for the two types of transformers are also vastly different. Transmission line transformers use chokes and transmission lines, while conventional transformers use flux linkages.

High-impedance ununs (and baluns), which match 50 ohms unbalanced to impedances as high as 800 ohms, lie at about the edge of this technology's capability. The reasons are: 1) the windings require more turns because higher reactances are needed for isolating the input from the output, and 2) they require higher characteristic impedances in the transmission lines because the loads they see are greater. Therefore, when winding one of these devices, you'll *just* run out of space on your toroidal cores when trying to satisfy the low frequency and high frequency objectives. Incidentally, beaded transmission lines are not recommended at these impedance levels because of their excessive losses.

There are two methods for obtaining broadband operation at these high impedance levels. One uses Guanella's 9:1 and 16:1 baluns, which are converted to unun operation.[2] The other uses higher-order windings (quadrifilar in this case) on a single core, which is an extension of Ruthroff's *boot-strap* approach.[2] The Guanella approach, which uses coiled transmission line connected in series at the high-impedance

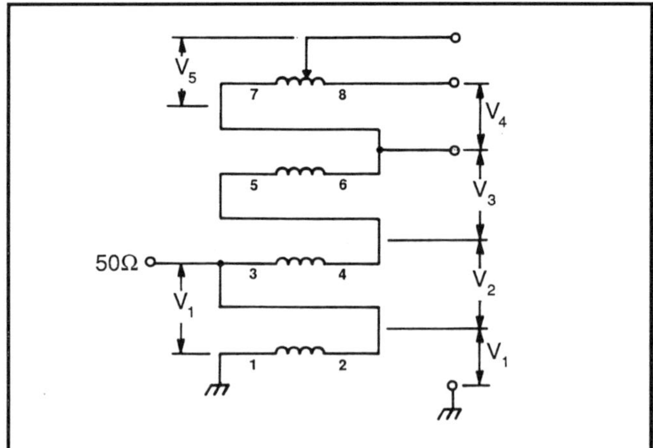

Figure 10-1. Schematic diagram of the quadrifilar design, using Ruthroff's approach, for high-impedance, low-frequency ununs like the Beverage matching transformer.

side and in parallel at the low-impedance side, results in very broad bandwidths—but with difficulty in meeting low-frequency objectives. Low-frequency models[2] show that, with ratios above 4:1, some of the coiled windings are connected in parallel, resulting in reduced reactances. However, with the Ruthroff approach, all of the inductances (at the low-frequency end) end up mutually aiding each other. However, Ruthroff's approach suffers at the high-frequency end because a direct voltage is summed with three voltages that traverse various lengths of transmission lines. As a result, Guanella's approach of summing voltages of equal delays is preferred for the higher frequency bands, and Ruthroff's approach is most often used for the lower frequency bands. This chapter presents designs using Ruthroff's approach.

Figure 10-1 shows the schematic diagram of a quadrifilar-wound unun. If the lengths of the transmission lines are very short compared to the wavelength (therefore, phase delay and standing waves are negligible), then:

$$V_1 = V_2 = V_3 = V_4 \qquad \text{(Eq 10-2)}$$

at terminal 6,

$$V_o = V_1 + V_2 + V_3 = 3V_1 \qquad \text{(Eq 10-3)}$$

and the impedance ratio becomes:

$$g = (V_o/V_1)^2 = 9 \qquad \text{(Eq 10-4)}$$

At terminal 8, it becomes:

$$g = 16 \qquad \text{(Eq 10-5)}$$

Photo 10-A. The bottom view of the low-power Beverage antenna unun.

The voltage at the tap in winding 7-8 is:

$$\begin{aligned} V_o &= 3V_1 + V_5 \\ &= 3V_1 + n/NV_1 = V_1(3 + n/N) \\ &= V_1(3 + n/N) \end{aligned} \qquad \text{(Eq 10-6)}$$

where:
N = total number of turns
n = number of turns from terminal 7

The impedance ratio, using the tapped winding, becomes:

$$g = (V_o/V_1)^2 = (3 + n/N)^2 \qquad \text{(Eq 10-7)}$$

When the lengths of the transmission lines are significant, then important phase delays can occur and reduce the high frequency response. As you can see in

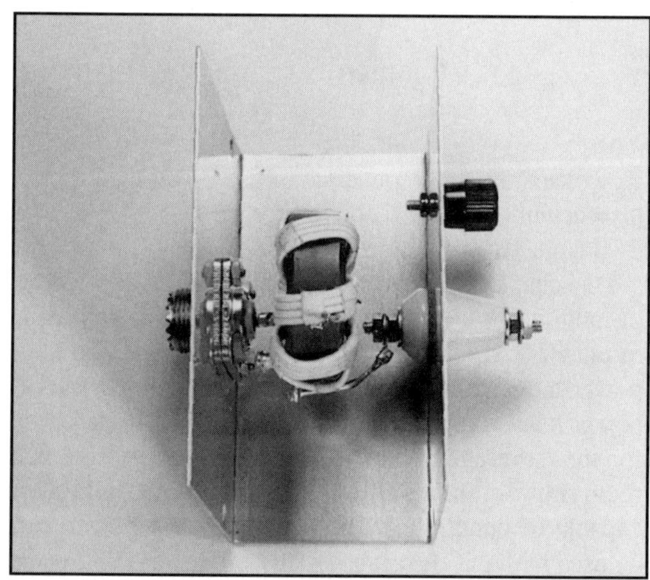

Photo 10-B. The low-power Beverage antenna unun mounted in a 4-inch long by 2-inch wide by 2.75-inch high minibox.

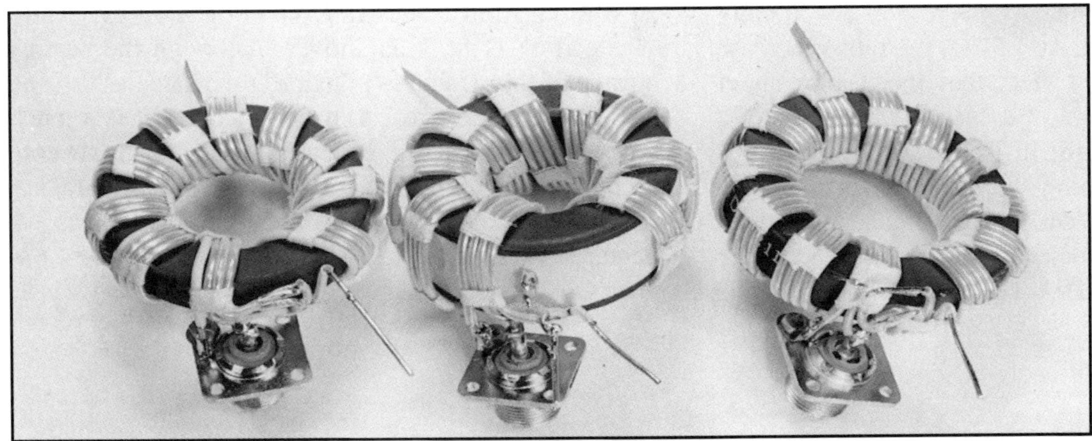

Photo 10-C. Three high-power, low-frequency ununs using a quadrifilar design with Ruthroff's approach. The one on the left is designed to cover the 80 and 160-meter bands. The other two are designed to cover the 160-meter and AM broadcast bands.

Figure 10-1, V_2 travels one transmission line, V_3 travels two transmission lines, and V_4 travels three transmission lines. Additionally, the high frequency response is further diminished if the characteristic impedances of the transmission lines are not at their optimum values (which is hard to do at these impedances levels). Even with these major flaws, the Ruthroff approach is better for Beverage antenna use because this antenna's greatest advantages are on the lower frequency bands (80 and 160 meters).

Sec 10.3 A Low-power Design

Photo 10-A shows the bottom view of a 5-turn, quadrifilar-wound unun designed to handle 100 watts of continuous power with constant ratios from 9:1 to 16:1 in the 40- and 80-meter bands. It uses the Ruthroff approach of **Figure 10-1** and is shown here to give the reader a method for making the various interconnections. For operation on the 80- and 160-meter bands, I would use 6 quadrifilar turns on a 1.5-inch OD ferrite toroid with a permeability of 250. The bottom winding is No. 20 hook-up wire and the other three are No. 22 hook-up wire. Winding 7-8 is tapped at 2 turns from terminal 7, yielding a 11.11:1 ratio, and at 3 turns from terminal 7, yielding a 12.25:1 ratio. Therefore, with outputs also at terminals 6 and 8, this unun matches 50-ohm cable to loads of 450, 555.6, 612.5, and 800 ohms.

Photo 10-B shows the unit mounted in a 4 inch long by 2 inch wide by 2.75 inch high minibox. The output (the feedthrough insulator) is connected to one of the taps. A grounded binding post is also shown.

Sec 10.4 High-power Designs

Photo 10-C shows three high-power designs. The one on the left is specifically designed to cover the frequencies generally used with traveling wave structures like the Beverage antenna. This design has 10 quadrifilar turns on a 2.4-inch OD ferrite toroid with a permeability of 250. Winding 1-2 is No. 14 tinned copper wire, and the other three are No. 16 tinned copper wire. The wires are also covered with Teflon™ sleeving. Winding 7-8 is tapped at 5 turns from terminal 7, yielding a ratio of 12.25:1. When matching 50-ohm cable to loads of 450 ohms (terminal 6), 612.5 ohms (the tap), and 800 ohms (terminal 8), the variation in response is less than 5 percent from 1.5 to 4 MHz. At 6.5 MHz, the variation (which is an increase in the impedance ratio) increases to about 20 percent. **Photo 10-D** attempts to provide a better view of the connections.

Photo 10-D. The bottom view of the high-power Beverage antenna unun.

The other two high-power ununs in **Photo 10-C** are specifically designed to cover the broadcast and 160-meter bands. The one in the center has 9 quadrifilar turns (of the same wires as above) on a stack of two 2.4-inch OD ferrite toroids with permeabilities of 250. The tap on winding 7-8 is now at 4 turns from terminal 7, yielding a ratio of 11.86:1. When matching 50-ohm cable to 450 ohms (terminal 6), 593 ohms (the tap), or 800 ohms (terminal 8), the response is literally flat from 0.5 to 2 MHz. At 4 MHz, the ratios increase by about 6 percent. At 7 MHz, they increase by about 20 percent.

The unun on the right in **Photo 10-C** illustrates another way of obtaining the same performance as above. In this case, the design has 12 quadrifilar turns (of the same wires as above) on a 2.68-inch OD ferrite with a permeability of 290. The tap on winding 7-8 is at 6 turns from terminal 7, yielding a 12.25:1 ratio (instead of 11.86:1 as above). Although the performance of this design is practically the same as the one above (using the two 2.4-inch OD cores), it is a much more expensive design because the 2.68-inch OD core is not nearly as popular. However, if a broadband, high-power and high-impedance unun (or balun) is required to cover 1.5 to 30 MHz, these expensive 2.68-inch OD ferrite cores are likely the *only* alternative!

Chapter 11

Concluding Remarks

Sec 11.1 Introduction

As I mentioned in the preface of this book, there is little information that deals with aspects of design and applications of practical hardware. Most companies are reluctant to publish their results for fear of giving away their hard-earned secrets. Textbooks only contain a few paragraphs on the subject. Therefore, very few people fully understand this technology and, as a result, it is still far from reaching its full potential.

This book not only contains my latest designs, which appeared in series written for *CQ* and *Communications Quarterly*, but also my views on recent articles that have appeared in the amateur radio literature.

In the process of converting and combining the articles into appropriate chapters, one section in my article in the *CQ*, March 1993 issue entitled "Dual-Ratio Ununs," stood out as having broader applications. It is entitled "Reflections on Power Ratings" and is presented here. This information is followed by a section on misconceptions and one on the "state of the art."

After reading this book, some might think I was overly critical and didn't agree with any of the designs or explanations (or both) presented in the amateur radio literature. This is quite true. In taking this stand I was hoping to provoke, in return, critical comment on my work. In this way, we can help our amateur friends—and perhaps even our professional friends—by advancing the understanding and application of these very useful transformers.

Sec 11.2 Reflections on Power Ratings

Power rating is one of the most controversial and least understood specifications for the transmission line transformer. As of today, no professional group has yet set the standard for this specification (as well as any other) for this popular class of matching transformers. In fact, manufacturers of ferrites, which are mainly the materials used with these devices, only specify them for their uses as conventional transformers and inductors, or microwave devices.

It is well known that power ratings for practically all conventional devices are based upon catastrophic failures (usually exceeding a voltage or current limit), and failures over a relatively short period of time due to an excessive rise in temperature.

With transmission line transformers, there are really two catastrophic-type failures that can occur. One is voltage breakdown. If the device is misterminated with a high impedance (especially an open circuit) a breakdown of the insulation can occur. This is particularly true of the 1:1 balun (50:50 ohms) that is terminated with the very high impedance of a full-wave dipole or inverted V. Using heavily coated wires (but still maintaining a characteristic impedance close to 50 ohms), or small but high-power coaxial cable, can help under these conditions.

The second catastrophic failure occurs at the low-frequency end of the transformer's passband, when the energy is not completely transmitted to the output circuit by a transmission line mode. This takes place when the reactance of the coiled or beaded transmission line is not sufficient to prevent conventional transformer currents or shunting currents to ground. Under these conditions, harmful flux can take place in the core or beads. Nonlinearities can also occur if the flux becomes appreciable. The objective in design at the low-frequency end is to have a margin of safety such that, with a termination of about three times (hence VSWR of 3:1) that of a matched condition,[2] no flux will appear in the core or beads.

The failure due to an excessive rise in temperature is the least understood of all because it involves the failure mechanism in transmission line transformers when only transmission line currents are allowed to flow. Unlike the conventional transformer whose losses are current dependent (wire, eddy current, and hystereis losses), the transmission line transformer's losses are voltage dependent (a dielectric-type). That is, the greater the voltage drop along the length of the transmission lines, the greater the loss. Furthermore, it can be shown that only low-permeability ferrites (less than 300) yield the extremely high efficiencies of which these transformers are capable.[2]

Because all transmission line transformers have voltage drops along their transmission lines, we must look at their high-frequency models to determine the magnitude of these drops and hence the temperature rise that can be expected. Here are some examples:

If V_1 is the voltage that appears on the 50-ohm side of the transformer, then:

1) For a quintufilar-wound unun, the longitudinal voltage-drop is $V_1/5$ in matching to lower impedances (like 32 or 18 ohms) and $V_1/4$ in matching to higher impedances (like 78 or 139 ohms).

2) For a quadrifilar-wound unun, the longitudinal voltage-drop is $V_1/4$ in matching to lower impedances (like 12.5 or 28 ohms) and $V_1/3$ in matching to higher impedances (like 89 or 200 ohms).

3) For a trifilar-wound unun, the longitudinal voltage-drop is $V_1/3$ in matching to lower impedances (like 22.22 or 25 ohms) and $V_1/2$ in matching to higher impedances like (100 or 112.5 ohms).

4) For a bifilar-wound unun, the longitudinal voltage-drop is $V_1/2$ in matching to a lower impedance of 12.5 ohms and V_1 in matching to a higher impedance of 200 ohms.

Because the quintufilar-wound unun has the lowest voltage drop, it is expected to have the highest efficiency. Furthermore, it can be seen that the highest efficiencies occur in matching to lower impedances. Very accurate measurements[2] have shown that 4:1 ununs, using ferrite cores with permeabilities of 125, have exhibited losses of only 0.02 to 0.04 dB in matching 50 to 12.5 ohms from 1 MHz to over 30 MHz. Even though the ununs in this book have mostly used permeabilities of 250, and should have slightly greater losses, many use higher-order windings (trifilar, quadrifilar, and quintufilar) and, hence, have lower longitudinal voltage drops. Therefore, they should have losses of only 0.02 to 0.04 dB, as well.

When matching at the 1 kW level, the figures above mean that only 5 to 10 watts would be dissipated in the unun. As a heatsink, these small transformers should be able to handle this loss easily. In fact, they should be able to handle several times this level of continuous power. Also, because they use heavily coated wires, their peak power ratings should be greater by more than a factor of two!

Another important power rating consideration is to determine what happens when the transformers are misterminated. Because the losses being considered now are dielectric-types and, hence, voltage-dependent, the harmful terminations are greater than that for which the transformers were designed. For example, if the termination is three times greater (a VSWR of 3:1), the voltages along the transmission lines would increase by a factor of 1.73. This means the losses would practically double. The ununs described in this series, when matching to impedances lower than 50 ohms, should easily handle this mismatch. Obviously, mismatches in the range of 10:1 would result in much lower efficiencies and should be avoided.

The analysis of the losses in baluns follows the same pattern. The voltages are as follows: under matched conditions for a 1:1 balun (50:50 ohms), $V_1/2$; for a 4:1 balun (50:200 ohms), V_1; for a 9:1 balun (50:450 ohms), $1.5V_1$; and for higher-impedance baluns it could be $2V_1$. The higher voltage drops, together with high VSWRs, means that high-impedance baluns (and ununs) have more loss and require larger structures to dissipate the heat. It should also be pointed out that there is a tradeoff in efficiency for low-frequency response with baluns (and ununs) when matching 50 ohms to higher impedances like 200 ohms, 300 ohms, 450 ohms, and higher. This is done by using permeabilities of 125 and lower.

Finally, I thought it might be useful to give some general guidelines as to what efficiencies you might expect with baluns and ununs when using ferrite cores or beads with a permeability of 250. Here are some expected efficiencies when matching 50 ohms to various loads under matched conditions:

Loads	Efficiency
50 ohms or less	98 to 99.5 percent
50 to 100 ohms	97 to 98 percent
100 to 200 ohms	96 to 97 percent
200 ohms and above	93 to 96 percent

As I mentioned earlier, these efficiencies would be reduced by a percent or two with a VSWR of 3:1, which increases the loss by a factor of about two. Also, the efficiencies can be increased by a percent or two with high-impedance loads (greater than 100 ohms) by resorting to lower permeability ferrites that trade off efficiency for low-frequency response. In closing, I would like to say that high-permeability manganese-zinc ferrites should be avoided because of their much higher losses. Furthermore, their losses are highly frequency dependent, while low-permeability nickel-zinc ferrites are not.

Sec 11.3 Misconceptions

From recent discussions "on-the-air" and phone calls concerning baluns, I think the most *expensive* misconception regarding baluns is the assumption that a 9:1 (450:50 ohm) balun would match 50-ohm cable (or the output of a linear or transceiver) to 450-ohm twin lead, without considering the effect of its termination. In truth, the 9:1 balun would *only* see 450 ohms if the line were terminated in 450 ohms. In reality, if the line were terminated in a 50-ohm dipole, the balun would see 50 ohms when the line is a half-wave long and **4050** ohms when is a quarter-wave long. The 9:1 balun is clearly useless in this application.

By far, most misconceptions regarding baluns are due to the many radio amateurs who perceive these devices as conventional transformers that transmit the energy from input to output by flux linkages and not as transmission line transformers, which transmit energy by an efficient transmission line mode. This is clearly shown by the writers who have compared their "new" coaxial cable (coiled about a toroid or threaded through ferrite beads) baluns with baluns using wire transmission lines coiled about a ferrite rod or toroid. They claim their baluns are better because the others: 1) were limited by leakage inductance, 2) did not exhibit true 1:1 impedance transformations, 3) were prone to core saturation, 4) added a reactive component to the input impedance, 5) were susceptible to unbalanced and mismatched loads, and more importantly, 6) had *more* loss.

If the writers had accepted the correct model for these devices (given to us by Guanella and Ruthroff), which shows that they are really chokes (lumped elements) and configurations of transmission lines (distributed elements), then there are several parameters they should have considered in their comparisons. They are: 1) the characteristic impedances and lengths of the transmission lines (the high-frequency capability), 2) what form of the 1:1 balun or 4:1 balun is used by the other balun, 3) the low-frequency capabilities (safety margins), 4) power capabilities, and finally 5) efficiencies.

Now, had the writers used the proper parameters in their comparisons, they would have found that mismatch loss was mistaken for real (ohmic) loss; high-frequency response was limited by standing waves, and not leakage inductance or shunting capacitance; the beaded-coax balun had *more* loss than a well-designed balun using wire or coax transmission lines coiled about a toroid; and that their comparisons were made with either the trifilar (voltage) 1:1 balun or the Ruthroff 4:1 balun, which are inferior designs.

In fact, the perception that the transmission line transformer is actually a conventional transformer is so prevalent, that a new name for this class of devices should be considered—*broadband transmission line matching networks*. This name (without the word *transformer*) would help in dispelling inaccurate perceptions and in standardizing the schematic diagrams. It would place the coiled or beaded transmission lines (in the high-frequency models) horizontally, and eliminate the phasing or polarity dots.

Sec 11.4 The State of the Art

Until very recently, the radio amateur had only two types of baluns available in the literature and on the market. They were the so-called 1:1 and 4:1 "voltage" baluns. As was shown in **Chapter 1**, the comparisons by others with new 1:1 designs using coaxial cable (called "current" baluns) were made with an inferior trifilar-wound balun, instead of Ruthoff's design that appeared in his 1959 paper and became the industry's standard. Ruthroff's third conductor on his 1:1 balun was on a separate part of the toroid, thus giving it practically the same characteristics as the Guanella ("current") balun. These recent articles on new designs not only gave a new language to our baluns, but also presented questionable statements regarding their performances. It would be interesting if the authors of these articles compared their baluns with well-designed Ruthroff or Guanella baluns using 50-ohm bifilar windings or coaxial cables on low-loss ferrite toroids (less than 300 permeability). I am quite sure their claims would be greatly diminished.

As was noted in Chapter 2, the 4:1 voltage balun appeared in the amateur radio journals about 25 years ago (the same time as the "inferior" 1:1 voltage balun). Considerable design information appeared in the handbooks of the time regarding the construction and performance of this balun. Furthermore, this information also stayed the same over these many years. As was shown in **Chapter 2**, the design was found lacking. However, with some rather simple changes, like doubling the cross-sectional area of the core, increasing the number of turns from 10 to 14, and using extra insulation on the wires to increase the characteristic impedance of the coiled transmission line from about 50 to 100 ohms (the objective), a much better design emerged. In fact, for balanced antenna systems, this new design might well be described as "peerless."

Lately, a 4:1 Guanella (current) balun appeared in our handbooks. This more flexible balun uses two transmission lines wound on separate cores and connected in series at one end and in parallel on the other. Literally, no design information is given on its construction. What is offered are recommendations for the permeability of the ferrite cores. Values from 850 to 2500 are proposed. However, use of these high permeabilities would result in lossy baluns.

I also found it interesting, in my work on these devices, that the classic papers of Guanella[3] and Ruthroff[9] are still the cornerstones of this technology known as *transmission line transformers*. To be sure, some of us have extended the work of these two by using better measuring equipment, creating more complicated configurations, and finding new applications. However, it is apparent from the articles published in the amateur radio journals and discussions "on-the-air" and at club meetings, that most radio amateurs still perceive these devices as conventional transformers. They don't look at these devices as Guanella and Ruthroff did—as chokes and transmission lines. As a result, there has been a lack of good design information in our literature.

This lack of good design information is not only endemic to the amateur radio literature. It also applies to the professional literature. Very little progress has been made in this field since Ruthroff's classic 1959 paper. From my vantage point, I see that the *transmission line transformer* technology has been literally frozen in time! However, there are many new and useful designs possible with this technology. They include: higher power levels, applications on the VHF and UHF bands and above, and new baluns and ununs with ratios other than $1{:}n^2$ where n = 1, 2, 3, . . . , etc. This book presents some designs* and suggestions for higher-power and higher-frequency applications.

I see two reasons for the lack of emergence in this technology. They are:

1) This subject is not adequately covered in any college textbook, and it certainly has not been of interest to academics who rightfully view their role as basic research and not applications. As a result, there are very few (if any) graduates with any skill in the design of transmission line transformers—in contrast to the areas of transmission line, waveguide, and antenna theory.

2) The professional societies don't receive enough application papers. Although much of the research and development work performed in industry is highly innovative, important to the advancement of the technology, and certainly publishable in scientific journals, corporations are often reluctant to allow publication for fear of "aiding" their competition. It has been stated[29] that in the last fifteen years, the submission of application papers to the technical journals of the IEEE has taken an *inexorable* slide. A recent survey by one of the technical societies showed that 85 percent of the submissions now come from universities, not industry!

In order to assist technologies, like *transmission line transformers*, which are far from reaching their potential applications, IEEE has instituted a new program called *Emerging Practices in Technology* (EPT). The object of the EPT program is to facilitate the development of new standards by disseminating and making available various EPT papers to the broadest possible audience worldwide. The papers on practices in various areas of technology are peer reviewed by relevant IEEE Technical Committees, and have the potential for standardization in the future. The papers (mine is **Reference 30**) are published by the IEEE Standards Press.

**Kits and finished units available from Amidon Associates, Inc., 2216 East Gladwick Street, Dominguez Hill, California 90220.*

Chapter 12

References

1. Jerry Sevick, W2FMI, *Transmission Line Transformers*, 1st edition, Amateur Radio Relay League, Newington, Connecticut, 1987.

2. Jerry Sevick, W2FMI, *Transmission Line Transformers*, 2nd edition, Amateur Radio Relay League, Newington, Connecticut, 1990.

3. G. Guanella, "Novel Matching Systems for High Frequencies," *Brown-Boveri Review*, Volume 31, September 1944, pages 327–329.

4. Walt Maxwell, W2DU, "Some Aspects of the Balun Problem," *QST*, March 1983, pages 38–40.

5. John Belrose, VE2CU, "Transforming the Balun," *QST*, June 1991, pages 30–33.

6. Roy Lewallen, W7EL, "Baluns: What They Do and How They Do It," *The ARRL Antenna Compendium*, Volume 1, Amateur Radio Relay League, Newington, Connecticut, 1985, pages 12–15.

7. Joe Reisert, W1JR, "Simple and Efficient Broadband Balun," *Ham Radio*, September 1978, pages 12– 15.

8. Richard H. Turrin, W2IMU, "Broad-Band Balun Transformers," *QST*, August 1964, pages 33–35.

9. C. L. Ruthroff, "Some Broad-Band Transformers," *Proceedings of the IRE*, Volume 47, August 1959, pages 1337–1342.

10. Richard A. Genaille, W4UW, "How to Build a Multi-Tap Unun," *CQ* May 1992, pages 28–32.

11. Bruce Eggers, WA9NEW, "An Analysis of the Balun," *QST*, April 1980, pages 19–21.

12. Richard H. Turrin, W2IMU, "Application of Broad-Band Balun Transformers," *QST*, April 1969, pages 42, 43.

13. George Badger, W6TC, "A New Class of Coaxial-Line Transformers, Part 1," *Ham Radio*, February 1980, pages 12–18.

14. George Badger, W6TC, "A New Class Of Coaxial-LIne Transformers, Part 2," *Ham Radio*, March 1980, pages 18–29,70.

15. Bill Orr, W6SAI, "Radio Fundamentals: The Coax Balun," *CQ*, November 1993, pages 60–65.

16. Albert Roehm, W2OBJ, "Some Additional Aspects of the Balun Problem," *The ARRL Antenna Compendium*, Volume 2, Amateur Radio Relay League, Newington, Connecticut, pages 172–174.

17. Lew McCoy, W1ICP, "Let's Talk About Wire," Part 1, *CQ*, January 1993, pages 42–45.

18. Lew McCoy, W1ICP, "Let's Talk About Wire," Part 2, *CQ*, February 1993, pages 32–42.

19. Gerald Hall, K1TD, Editor, *The ARRL Antenna Book*, 16th edition, (Newington, CT: ARRL, 1991), Chapter 25.

20. George Grammer, W1DF, "Simplified Design of Impedance-Matching Networks," in three parts, *QST*, March, April, and May 1957.

21. D.K. Belcher, "RF Matching Techniques, Design and Example," *QST*, October 1972, page 24.

22. Lew McCoy, W1ICP, "The Ultimate Transmatch," *QST*, July 1970, pages 24–27, 58.

23. Warren Bruene, W5OLY, "Introducing the Series-Parallel Network," *QST*, June 1986, page 21.

24. Louis Varney, G5RV, "The G5RV Muliband Antenna . . . Up-to-Date," *The Antenna Compendium*, Volume 1, Amateur Radio Relay League, Newington, Connecticut, pages 86–90.

25. Jerry Sevick, W2FMI, "6:1 and 9:1 Baluns," *Communications Quarterly*, Winter 1993, pages 43–51.

26. D. Meyer, "Equal-Delay Networks Match Impedances Over Wide Bandwidths," *MICROWAVES & RF*, April 1990, pages 179–188.

27. S.E. London and S. V. Tomeshevich, "Line Transformers with Fractional Transformation Factor," *Telecommunications and Radio Engineering*, Volume 28/29, April 1974.

28. Carl Markle, K8IHQ, "Wideband RF Baluns," *73 Amateur Radio Today*, September, 1992, pages 20–23.

29. S. Maas, D. Hornbuckle, D. Masse, "Applications Papers for the *MTT Transactions*," Summer 1992, IEEE MTT-S Newsletter, pages 3, 5.

30. Jerry Sevick, W2FMI, "Design and Realization of Broadband Transmission Line Matching Transformer," *Emerging Practices in Technology*, IEEE Standards Press, 1993.

Appendix I

Ground-Radial System for Verticals

A job transfer in 1970, and a subsequent move to a new neighborhood, prompted me to look at my antenna needs in a new light. I wanted to minimize, or at best, avoid any obstacles to good neighborly relations by not immediately reinstalling my triband Yagi beam on its 40-foot tower. A ground-mounted vertical beam antenna, i.e., an array of vertical antennas coupled together with the appropriate feed system, seemed the logical choice for my new installation because it would:

1) Exhibit a low profile.
2) Not require extra help in installation.
3) Offer a low angle of directed radiation.
4) Not require a large outlay of money.
5) Give me some experience with vertical antennas.

Consequently, I constructed a square array of four vertical antennas with four quarter-wave radials (on the surface of the earth) under each vertical. When my tests began, it was immediately apparent that the simple procedures I was using were inadequate to cope with such a complex system. I had to start anew to develop a test procedure and build suitable test equipment. Because I didn't want to dig up my backyard for buried radials, and no practical information was available for radials on the earth's surface, the logical step was to backtrack to a single vertical antenna and

Photo I-A. The experimental setup I eventually used in my vertical (and other ground-fed) antenna studies. It has a large aluminum plate (10 to 15 inches in diameter), a resistive bridge with external meter (Chapter 12, Reference 2), and a variable signal source. In this case, the transceiver in the shack is the signal source.

use it as a model upon which to develop some basic standards. I found that a simple impedance bridge (**Chapter 12, Reference 2**), a field-strength meter, and a test oscillator (**Chapter 12, Reference 2**) provided me with all of the data about verticals I needed. The results presented in this appendix are largely taken from my July 1970 *QST* article "The Ground-Image Vertical Antenna."

Ground versus Elevated Radials

A ground-image antenna differs considerably from a ground-plane antenna that relies merely on a few quarter-wave radials above the ground. The ground-image antenna results when an adequate number of radials are used, so the image of only the vertical section is sufficient to describe it. Practical information on ground-image systems has appeared in the literature for many years.[1,2] The results have shown that some 100 half-wave radials buried just below the surface of the earth provide a good ground system. However at higher frequencies, the dielectric effect of the earth brings about severe discrimination of radiation or reception at very low angles.[3,4]

The ground-plane vertical, which uses quarter-wave radials (or odd-multiples thereof), can shield the vertical portion from the lossy earth depending upon the number of radials and the height above ground. It is apparent that the higher the antenna, the fewer the number of radials needed. The minimum number is three radials spaced 120 degrees apart. The disadvantage of this type of antenna at low frequencies (1.7 to 7.3 MHz) simply lies in the difficulty of constructing a system that has a sufficient number of radials and enough height to be an efficient radiator. Furthermore, if the installation is in your backyard, you might have a problem cutting the lawn!

For these reasons, I chose to investigate the effect of radials *just on* the surface of the earth and at frequencies greater than 3 MHz. The results of my findings were most gratifying. It became apparent that many "rules of thumb" which have been taken for granted over these many years were more myth than truth.

Experimental Results

The experimental setup I eventually used in my vertical (and other ground-fed) antenna investigations is shown in **Photo I-A**. It has a large aluminum plate (10 to 15 inches in diameter) to which the radials are connected, a resistive bridge with an external meter,

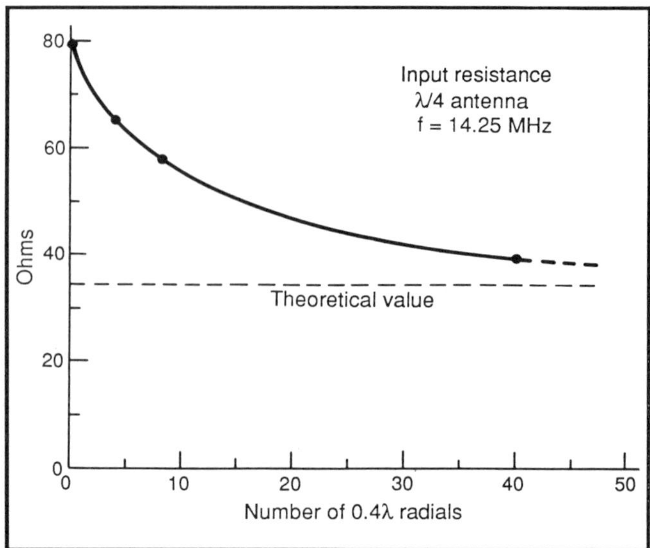

Figure I-1. The resonant input impedance of a 20-meter quarter-wave vertical as a function of the number of 0.4 wavelength long radials just on the surface of the earth. The 4 and 8 radials consisted of bundles of 5 No. 18 copper wires. The 40-radial point was obtained by fanning out the 8 radials.

(**Chapter 12, Reference 2**) and a variable signal source. In the example shown, the signal source was the transceiver in my shack. Reading the classic paper by Brown, Lewis, and Epstein,[1] it appeared that a good starting point was to use radials 0.4 wavelengths long on the 20-meter band. Each radial, in turn, was terminated by large nails (10 to 12 inches long). These nails not only secured the radial ends, they also provided an additional electrical contact with the earth.

The parameter I chose to measure was the resonant input impedance of a quarter-wave vertical antenna. It is known to be:

$$R_{in} = R_{rad} + R_{loss} \quad \text{(Eq I-1)}$$

where:

R_{in} = the resonant input impedance
R_{rad} = the radiation resistance
R_{loss} = the loss in the ground system

In order to obtain some idea of the number of radials required for an adequate ground system, I used eight bundles of wire, each 25 feet long, and each made up of five No. 18 copper wires. I was then able to use each bundle as a radial, measure the input impedance, and then finally separate the wires in the bundles to arrive at 40 radials.

Figure I-1 shows the resonant input impedance as a function of the number of radials. The results are surprisingly close to those of Brown, Lewis, and Epstein,[1] which were obtained with buried radials in the AM broadcast band. My technique points out some important features and refutes some of the old myths of ground systems:

1) Radials on the surface of the ground are as effective as buried radials. In fact, the radials should not be buried too deeply because the electric field only penetrates the earth a few feet at HF.

2) Four radials (even 0.4 wavelengths long) make a *very* poor ground system.

3) Forty radials were required at my location in order to provide an adequate ground system. Later work showed that 0.2 wavelength radials provided the same result (see **Appendix II**).

4) Because each radial carries only 1/*n*th of the current, where *n* is the number of radials, and practically any metallic conductor is much better than the lossy earth, the type of conductor (copper, aluminum, or iron) and size is not very important. The main criteria is how long the radials will survive in the environment. This suggests that insulated wires could be the best way to go.

Another myth, which is frequently heard "on-the-air," is the idea that the minimum VSWR always occurs at the resonant frequency of an antenna. In fact, just the opposite is true. **Figure I-1** shows the radiation resistance of my vertical, which has a height-to-radius ratio of 300,[5] is 35 ohms. Therefore, over a perfect ground system, the VSWR would be 1.4:1 at resonance. However, one finds that a minimum VSWR of 1.3:1 occurs a little above the resonant frequency. Even though the input impedance takes on an inductive component above resonance, the real part (the radiation resistance) also increases—resulting in the lower VSWR. Another misconception is that a 1:1 VSWR with a ground-mounted quarter-wave vertical is an ideal condition. Again, not true. It shows that the ground-system loss is on the order of 15 ohms. With shortened verticals, the loss could even be greater.

Another aspect of this study was my investigation of the effect on the low-angle radiation as a function of the number and length of the radials. **Figure I-2** shows the improvement in the low-angle radiation as a function of the number of 0.4-wavelength radials. I obtained these data by placing a test oscillator on a 26-foot wooden tower 4 wavelengths away from the 20-meter vertical.

The effect of using longer radials in a particular direction is shown in **Figure I-3**. Radials of No. 18 wire 3/2 wavelengths long were placed between the existing 40 radials. The spacing between the longer wires was 5 degrees. The considerable improvement indicates the need for longer radials and the directional properties a ground plane could provide.

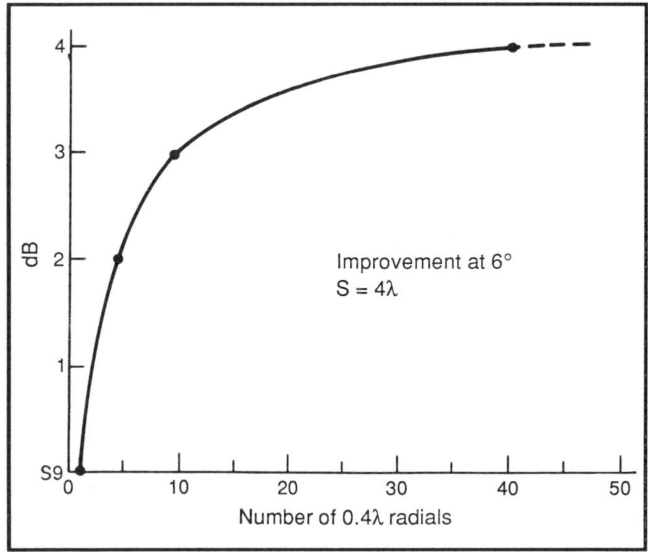

Figure I-2. The improvement in low-angle radiation of a quarter-wave vertical on 20 meters as a function of the number of added 0.4 wavelength radials. A test oscillator was mounted on a wooden tower four wavelengths away at an elevation angle 6 degrees from the base of the vertical.

I also built and tested a 5/8-wavelength vertical. It consisted of a 40-foot telescoping aluminum pole and a loading coil of 8 turns of No. 12 wire with a diameter of 2.5 inches. A field strength meter was mounted at different heights on the 26-ft tower. The comparison in low-angle radiation at a distance of 7.5 wavelengths is presented in **Table I-1**. The ground plane for these data consisted of the forty 0.4-wavelength radials plus the eleven 3/2-wavelength ones. Measurements were taken in the direction of the added longer radials. The results show the improvement in low-angle radiation offered by the 5/8-wavelength vertical. The resonant input impedance of this much taller vertical was found to be 76 ohms.

On-The-Air Checks

Taking these experiments and measurements into account, I settled on a 1/4-wavelength vertical with forty 0.4-wavelength (unburied) radials. I then proceeded to make on-the-air tests to compare the effectiveness with an inverted-V antenna having its apex at 0.4 wavelength, and with a 5/8-wavelength vertical over the same ground plane. Surprisingly, the 1/4-wavelength vertical seemed to perform just as well as

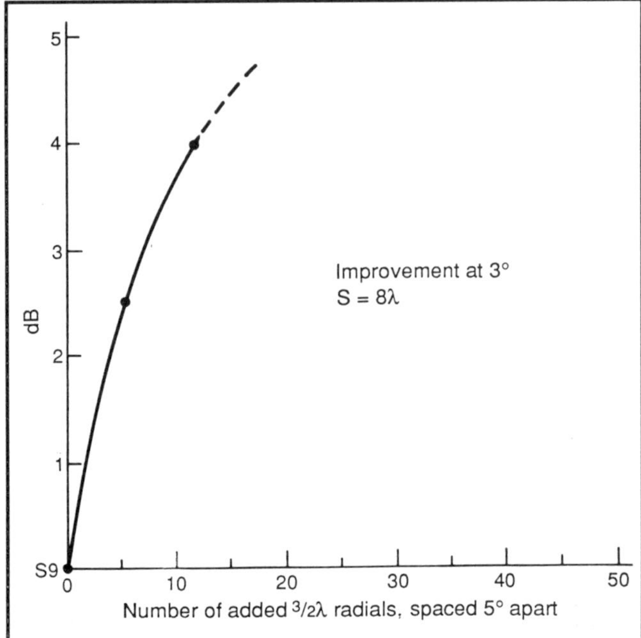

Figure I-3. The result of interlacing 3/2-wavelength radials in a particular direction. Data were obtained by using a test oscillator on a wooden tower eight wavelengths away, at an elevation of 3 degrees from the base of the vertical. The results indicate the advantage of longer radials and the possibility of the directional properties of a nonsymmetrical radial system.

θ	E(λ/4)	E(5/8λ)	Gain of 5/8λ antenna
0.1 degree	0	0.58	∞
0.4 degree	0	0.62	∞
0.75 degree	0.62	0.80	3.4 dB
1.1 degree	0.69	1.0	3.2 dB
1.5 degrees	0.62	0.92	3.4 dB
2.25 degrees	0.48	0.80	4.3 dB
3 degrees	0.41	0.69	4.5 dB

Table I-1. Comparison of responses of 1/4-wavelength and 5/8-wavelength vertical antennas at low radiation angles. Data were taken by a field-strength meter mounted on a wooden tower at a distance of 7.5 wavelengths and at 14.250 MHz. Field strength, E, is normalized to the maximum value obtained with the 5/8-wavelength case.

the much taller 5/8-wavelength vertical. This could result from the fact that most signals arrive after several hops, and the optimum lobe angle is probably as high as 15 to 20 degrees.[6] At that angle, the 1/4-wavelength vertical actually enjoys an advantage. With practically all DX contacts, the verticals had a 6- to 8-dB improvement over the inverted-V! The only exceptions were at intermediate distances and for local contacts. At about 500 to 600 miles, the inverted-V with its higher angle of radiation gave better results. Locally, the verticals gave far superior performances. Improvements of 10 to 15 dB were recorded. I also tested a triband trap-vertical antenna on 20 meters and found to be practically the same in impedance and performance as the 1/4-wavelength vertical. This antenna had an overall height of only 12.5 feet!

Closing Comments

The verticals in this study were matched to 50-ohm cable using L-C networks. At this time, I was only able to construct 4:1 ununs matching 50-ohm cable to unbalanced loads of 12.5 ohms. However, as is noted in **Chapter 7**, simple fractional-ratio ununs are now available to match 50-ohm cable to the impedances of the verticals in this study—namely, 35 and 76 ohms.

Finally, it should be mentioned that the world standard for the number of radials to be used with verticals in the AM broadcast band is 120. This number was based on the classic paper published in 1937 by Brown, Lewis, and Epstein. During the course of a business meeting with Dr. Brown, I asked him how he and his colleagues arrived at the 120 radial figure—because I was quite sure 100 would work as well. His answer was interesting.

He said that he and the others had been thinking in terms of 100 radials, but the farmer who plowed in 100 radials had wire left over because copper is soft and stretches easily. When he asked what to do with the extra wire, the farmer was told to plow it in. The result was a world standard of 120 radials.

References

1. Brown, Lewis, and Epstein, "Ground Systems as a Factor in Antenna Efficiency," *Proceedings of the IRE*, Vol. 25, No. 6, June 1937.

2. Wait, "Input Resistance of L. F. Unipole Aerials," *Wireless Engineer*, May 1955.

3. Feldman, "The Optical Behavior of the Ground for Short Radio Waves," *Proceedings of the IRE*, Vol. 21, No. 6, June 1933.

4. Jager, "Effect of the Earth's Surface on Antennas Patterns in the Short Wave Range," *Internet Elekitron Rundschau 1970*, Nr. 4.

5. King and Harrison, "The Impedance of Short, Long, and Capacitively Loaded Antennas with a Critical Discussion of the Antenna Problem," *Journal of Applied Physics*, Vol. 15, February 1944.

6. Friis, Feldman, and Sharpless, "The Determination of the Arrival of Short Radio Waves," *Proceedings of the IRE*, Vol. 22, No. 1, January 1934.

Appendix II

Short Ground-Mounted Verticals

Appendix I reviewed the highlights of my first *QST* article, which was published in July 1971, and was entitled "The Ground-Image Vertical Antenna." From what I learned about ground-radials on the surface of the earth, I was able to construct a low-loss ground system for a parasitic vertical beam (actually one-half of a Yagi beam) that proved to be very competitive. This antenna was described in the June 1972 issue of *QST* as "The W2FMI 20-Meter Vertical Beam."

In the process of trying to extend the results of these two investigations to the 40-, 80- and 160-meter bands, and to possible multiband use, the need arose to understand the operation of a shortened vertical and the effects of different loading schemes on the input impedance. I certainly was not interested in full-sized quarter-wave verticals especially on 80 and 160 meters. The results of this third investigation were published in the March 1973 issue of *QST* in an article entitled "The W2FMI Ground-Mounted Short Vertical." This appendix highlights some of the more important results presented in that article.

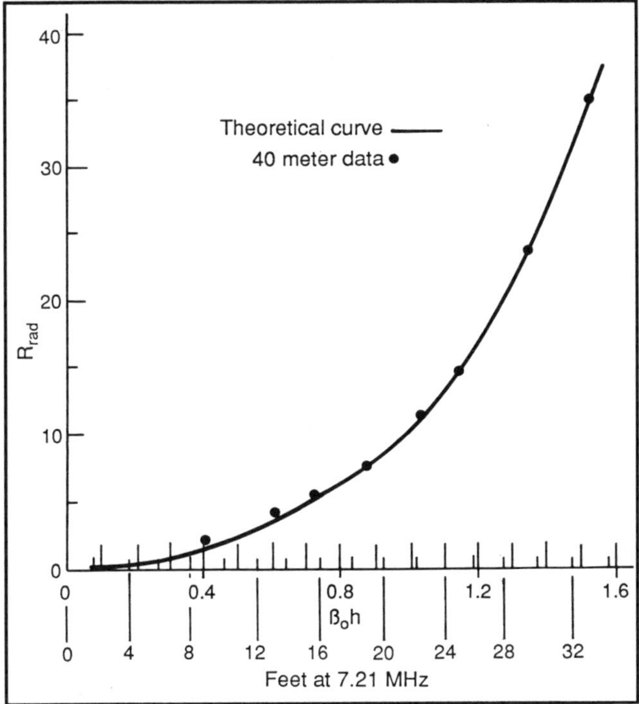

Figure II-1. Theoretical curve and experimental results for the radiation resistance as a function of height.

The first part of **Appendix II** deals with the theoretical considerations of a short vertical antenna. This is followed by experimental results that show the trade-offs involved in shortening antennas by various schemes. Finally, specific designs are given for the 40- and 80-meter bands. As you will see, a short vertical antenna, properly designed and installed, approaches the efficiency of a full-size quarter-wave vertical antenna. Even a 6-foot vertical on 40-meters can produce an exceptional signal. Moreover, as this book has shown, efficient and broadband ununs are now available to match 50-ohm cable to practically any short vertical antenna.

Theoretical Considerations

The short antenna has been defined as one that is small compared to a wavelength. In a more exact form, it is defined in such a manner as to simplify the mathematics in the theoretical calculations. King[1] has used the following inequality as the definition

$$\beta_o h \leq 0.5 \qquad \text{(Eq II-1)}$$

where:

$\beta_o = 2\,\pi/\lambda$

h = half-length of a center-fed dipole or the height of a ground-mounted vertical

λ = the wavelength

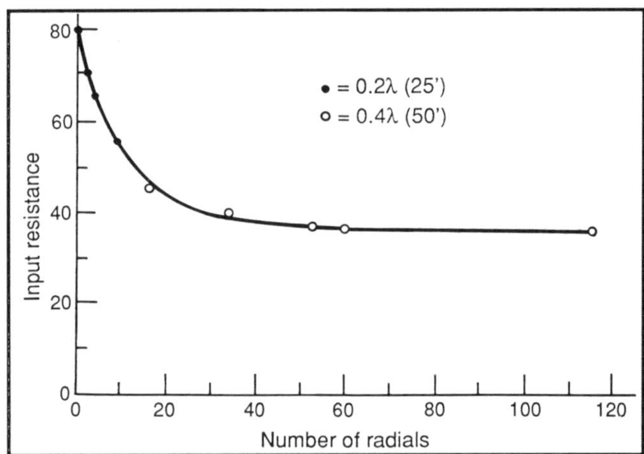

Figure II-2. Input resistance of a quarter-wave ground-mounted resonant vertical as a function of the number of radials.

$\beta_o h$ is actually a quantity used to express the height of an antenna in terms of an angle in radians. Thus, this quantity is independent of the frequency. Because I used 40 meters most for the experimental results presented in this appendix, the inequality above assures accurate theoretical calculations for verticals of eleven feet or less.

The theoretical results show that the power gain (when compared to an isotropic radiator) for a *very* short antenna, even one that is less than one foot high on 40 meters, is 1.5.[1] This increases slowly to 1.513 for an eleven-foot vertical. These gains are to be compared to about 1.62 for a resonant quarter-wave vertical. This difference amounts to less than 0.4 dB or 0.07 S unit, based on 6 dB per S unit.

In a similar fashion, the related parameter known as the capture cross-section also differs by relatively small amounts. For the very short antenna, a cross-section value is $0.119\lambda^2$ while for the quarter-wave vertical it is $0.13\lambda^2$. This is surprising to many because it is difficult to visualize that a vertical of a foot or two in height on 40 meters could have practically the same receiving ability as a 33-foot quarter-wave antenna!

The *very small value of its input resistance* is the important property of a short vertical that makes its capture cross-section nearly the equivalent of a full quarter-wave vertical. **Figure II-1** shows the theoretical curve,[1] and the 40-meter experimental results for the input resistance of a ground-mounted vertical as a function of height. The experimental data were obtained by essentially canceling out the reactance of the shortened vertical by placing an inductance at its base and measuring the resistive value with a simple impedance bridge. Because I used an extensive ground system of 115 radials of No. 15 aluminum wire each 0.4 wavelengths long, together with base loading coils with Qs approaching 900 (see **Appendix IV**), the resistance measured was actually that of the antenna itself. This is called the *radiation resistance*.

Figure II-2 shows the input resistance of a quarter-wave resonant vertical as a function of the number of radials on the earth's surface and their termination in large nails (10 to 12 inches long). At the 115-radial point, the input resistance approaches the theoretical value of 35 ohms, which strongly indicates low earth loss and reliable data in short vertical antenna measurements. **Photo I-A** in **Appendix I** shows the experimental setup. Even though I used No.15 aluminum, I am quite sure that No. 22 or even No. 28 copper wire would work just as well.

It should also be noted from **Figure II-2** that I noticed very little difference whether 0.2 or 0.4 wavelength radials were used. The results are practically identical with sixteen radials. In fact, this curve is also similar to the one shown in **Appendix I. Appendix III** will show some results of the trade-offs in performance as a function of the lengths of the (unburied) radials.

Experimental Results

In designing a shortened vertical for 40, 80 or 160 meters, it is most important to obtain the highest possible resonant input impedance depending upon the mechanical constraints. It is a well-know fact that base-loading, shown in **Figure II-1**, yields the lowest value. Because data on other forms of loading—i.e., top-hat, three-quarter point, midpoint, and distributed (helical antenna)—were not readily available, I undertook an experimental investigation to see what increases could be obtained in the radiation resistance. **Photo II-A** shows some of the different forms of loading used in this investigation.

The results of my experiments are shown in **Figure II-3. Table II-1** shows the individual points for inductive loading and **Table II-2** for top-hat and distributed loading. These experiments brought out several interesting results. According to **Figure II-3**, top-hat loading yielded the largest value of radiation resistance for a particular height. A top hat is also considered the most low-Q loading element. Surprisingly, the helical antenna[3] yielded a value less than midpoint loading. The three-quarter point and midpoint loading curves

Photo II-A. Some of the different forms of loading that were used in this investigation.

were not extended to lower values of height because data were very difficult to obtain below the points shown on the respective curves. The combinations of inductances and lengths below the heights shown on these two curves were probably beyond resonant conditions at the frequency used in the measurements. I extended the other curves with dashed lines indicating no difficulties were encountered in the measurement, and therefore other lengths were very possible.

Several other interesting comments can be made about top-hat loading. **Table II-2** shows that the reduction in height due to top-hat loading (with a conducting wire around the perimeter) is approximately equal to twice the diameter of the top hat. Also, a four-spoked wheel approximates, to a good degree, a solid disk. Doubling to eight spokes only improves the loading by about 9 percent. Thus, *a few radials on the top of a vertical, which are electrically connected by a perimeter conductor, are very effective*. Four radials at the base, approximating a ground system, are practically *useless* as noted in **Figure II-2**.

Top hats have been made successfully on 160 meters with sloping wires or struts.[2] Because of the sloping nature of the top hat, some canceling of the radiation occurs—reducing the radiation resistance further.

Although the curves in **Figure II-3** were obtained from experiments at 7.21 MHz, these data can be applied to other bands by proper scaling. For example, by doubling all dimensions, including numbers of turns of the loading coils, the radiation resistance values would apply at a frequency of 3.6 MHz. By increasing the dimensions by 1.85 instead of 2, the results would apply at a frequency of 3.9 MHz. In like manner, a proper scaling factor could be used to apply these results to a portion of any of the amateur bands.

40- and 80-Meter Short Vertical Designs

The main objective in good short vertical antenna design is to have the radiation resistance much greater than the ohmic losses in the loading coils and ground systems. Coils with Qs approaching 900 (see **Appendix IV**) minimize the problem with loading coil losses. A sufficient number and length of ground-radials (for a ground-mounted vertical) eliminates the issue with ground losses. Because I used high-Q coils and 115 radials, any radiation resistance above a few

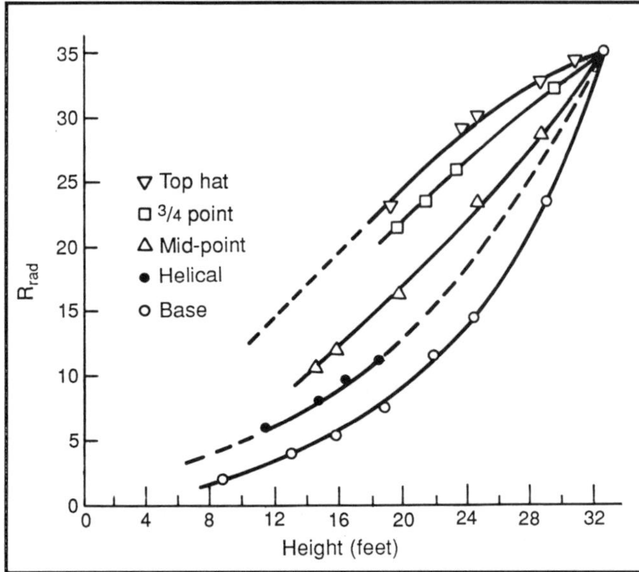

Figure II-3. Experimental results of radiation resistance as a function of height of antenna for various types of loading (40-meter data).

ohms gave me the opportunity to verify the theoretical predictions for very short verticals and to present designs that should be easily reproduced.

Because a broadband 4:1 unun was available from my previous *QST* article on a 20-meter parasitic beam, the first design was for a short vertical having a resonant input impedance close to 12.5 ohms. The curves of **Figure II-3** indicated that a 16- or 17-foot antenna with a coil of some 13 turns at 8 or 9 feet from the base would provide the proper impedance on 40 meters. I decided that some 7 to 8 feet of length above the coil could be replaced with a 4-foot top hat. The actual design that resulted had a total height of 10 feet and a 14-turn coil mounted one foot below the top hat. This height was about one foot higher than first expected; but upon careful examination, I found that the top hat also lowered the radiation by replacing a section of the vertical portion. Therefore, the height had to be increased some in order to counteract the top-loading effect.

I investigated several other shortened 40-meter vertical antennas. One was an 8.5-foot helical vertical using a 4-foot top hat and 75 turns of wire on a 1.625-inch, 7 foot long wooden dowel. The input impedance was 7.5 ohms and it was matched to 50-ohm cable with a standard pi network. Several tests were made by doubling the winding pitch below and above the midpoint of the dowel, keeping the number of turns constant. There was very little difference in the results.

1) Base Loading

No. of Turns*	h	$R_{rad.}$
a) 7	24 feet 3 inches	14.5 ohms
b) 10	18 feet 9 inches	7.5 ohms
c) 12	15 feet 7 inches	5.5 ohms
d) 14	13 feet	4 ohms
e) 18	8 feet 10 inches	2 ohms

2) Midpoint Loading

No. of Turns*	h	$R_{rad.}$
a) 6	28 feet 5 inches	28.5 ohms
b) 11	24 feet 3 inches	25.2 ohms
c) 15	19 feet 8 inches	16.5 ohms
d) 18	15 feet 8 inches	12.3 ohms
e) 24	14 feet 6 inches	10.5 ohms

3) Three-Quarter Point Loading

No. of turns*	h	$R_{rad.}$
a) 10	29 feet 4 inches	32.5 ohms
b) 18	23 feet	26 ohms
c) 23	21 feet 2 inches	23.5 ohms
d) 24	19 feet 2 inches	22 ohms

**B&W 3029, 2-1/2 inch diameter, 6TPI, No. 12 wire.*

Table II-1. Inductive loading.

1) Top Hat Loading (4-spoked wheel with 1/8-inch aluminum wire rim)

Diameter	h	$R_{rad.}$
a) 1 foot	30 feet 10 inches	34 ohms
b) 2 feet	28 feet 7 inches	32.5 ohms
c) 4 feet	24 feet	30 ohms
d) 7 feet	19 feet 2 inches	23.5 ohms
e) 4 feet*	23 feet 4 inches	29.4 ohms

2)Distributed Loading (helical antenna)

No. of turns	h	$R_{rad.}$	Top Hat at 1.5 feet Above Coil
a) 111	12 feet	8 ohms	1 foot dia.
b) 105	12 feet	10 ohms	2 feet dia.
c) 113	7 feet	6 ohms	2 feet dia.
d) 75	7 feet	7.5 ohms	4 feet dia.

**8-spoked wheel.*

Table II-2. Top Hat and distributed loading.

The next design, using a 7-foot top hat and a 14-turn loading coil 6 inches below it, resulted in a resonant vertical only 6 feet high. The input impedance was only 3.5 ohms. Because an unun with an impedance ratio of near 16:1 was not available at this time, matching was accomplished with a 4:1 unun and a pi network. **Photo II-B** (which was taken in 1972) shows the author and the 6-foot vertical.

I also investigated a larger 40-meter vertical. It had a 4-foot top hat and 7 turns of base loading that result-

Photo II-B. The author with the 6-foot, 40-meter vertical.

Total Height	$R_{rad.}$ (ohms)	No. turns*	Diameter Top (feet)	Bandwidth
6 ft	3.5	14 at 6 inches below Top Hat	7	100 kHz
8-1/2 ft.	7.5	75 on 7-foot dowel 1-8/5 inches in diameter	4	100 kHz
10 ft.	12.5	14 one foot below Top Hat	4	125 kHz
15 ft. 10 in.	12.5	7 at base	4	540 kHz

**Except for helical antenna, coil wire is same as shown in Table II-1.*

Table II-3. Parameters of 40-meter short vertical designs.

ed in a height of 15 feet, 10 inches (approximately a 1/8 wavelength) and an input impedance of 12.5 ohms. The purpose of this design was to compare its low-angle radiation and bandwidth (defined as the range in frequency where the VSWR is less than 2:1) with the other antennas. **Table II-3** shows the parameters of these various short verticals. **Figure II-4** shows the VSWR curves of three of the short verticals compared to a quarter-wave vertical and **Figure II-5** shows the VSWR curve of the 6-foot antenna.

As the VSWR curves indicate, shortening an antenna generally decreases its bandwidth. Also, if one compares the 8.5-foot and 6-foot verticals, top-hat loading appears to affect the bandwidth the least. Note that the 1/8-wavelength vertical appears to have a reasonable bandwidth. This could result in a practical design for a vertical beam that incorporates several 1/8-wavelength verticals.

Next, I compared the four short-vertical designs described above with a 1/4-wavelength vertical for low-angle radiation and performed on-the-air checks with other amateur signals. The low-angle radiation measurements were made three wavelengths in distance and at heights of 1, 3, 6, and 8 feet. This relates to vertical angles of 0.15, 0.45, 0.90, and 1.2 degrees. All measurements were made under matched conditions, with a constant 100 watts into the antenna under test. In no case were there any appreciable differences noted in the field strength measurements. In fact, the

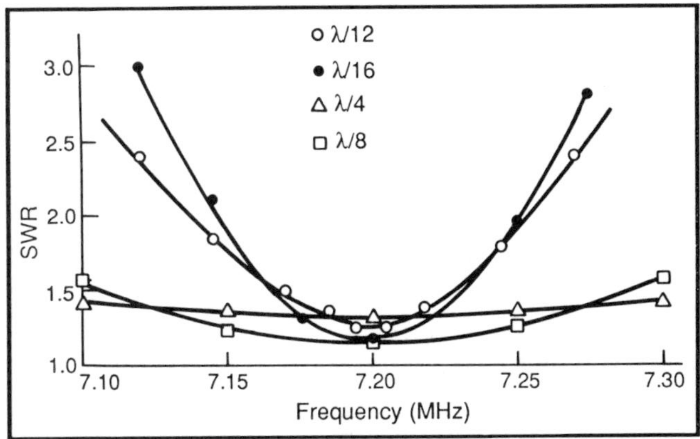

Figure II-4. Standing wave ratio of various short verticals compared to a quarter-wave antenna. The 1/16-wavelength antenna is the 8-1/2-foot helical design, the 1/12-wavelength design is the 10-foot antenna, and the 1/8-wavelength one is the 15-foot, 10-inch antenna. All three are described in the text.

6-foot vertical seemed to give slightly higher readings. These measurements certainly tend to verify the theory on the power gain predicted for short verticals. On-the-air checks were again very gratifying and exciting. Over two hundred contacts with the 6-foot antenna strongly indicated the efficiency and capability of a short vertical. Invariably, at distances greater than 500 to 600 miles, the short verticals yielded excellent signals.

After I had completed my investigation of short 40-meter verticals, I chanced upon a beach umbrella in a sports store. This umbrella had dimensions close to the 6-foot antenna used in my study. Without a second

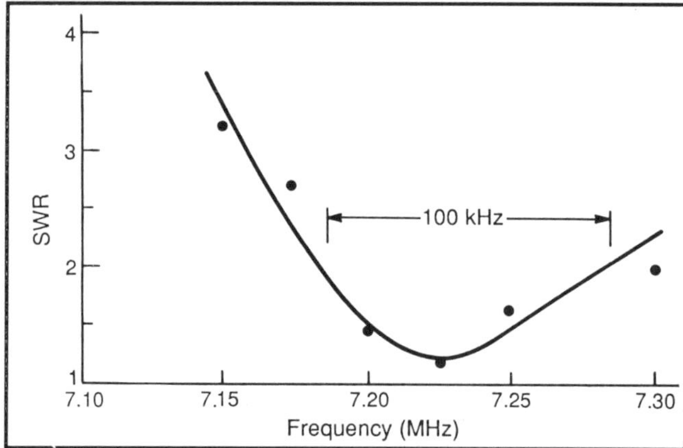

Figure II-5. Standing wave ratio of the 6-foot vertical using a 7-foot top hat and 14 turns of loading six inches below the top hat.

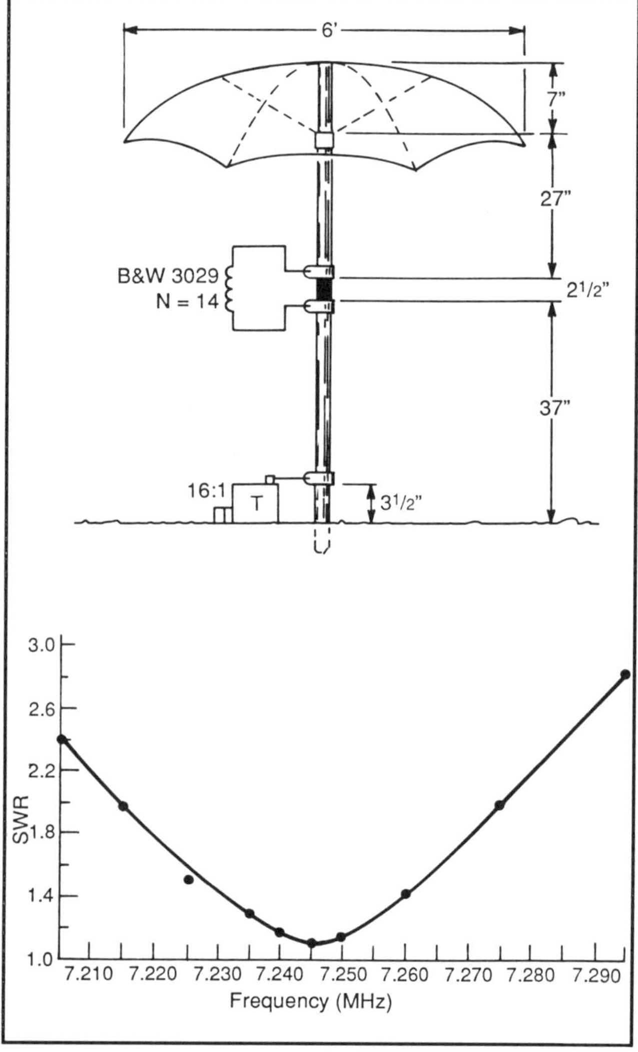

Figure II-6. The parameters and standing wave ratio as a function of frequency for the W2FMI 40-meter beach umbrella antenna.

Photo II-C. The author's XYL sitting under the W2FMI 40-meter beach umbrella antenna.

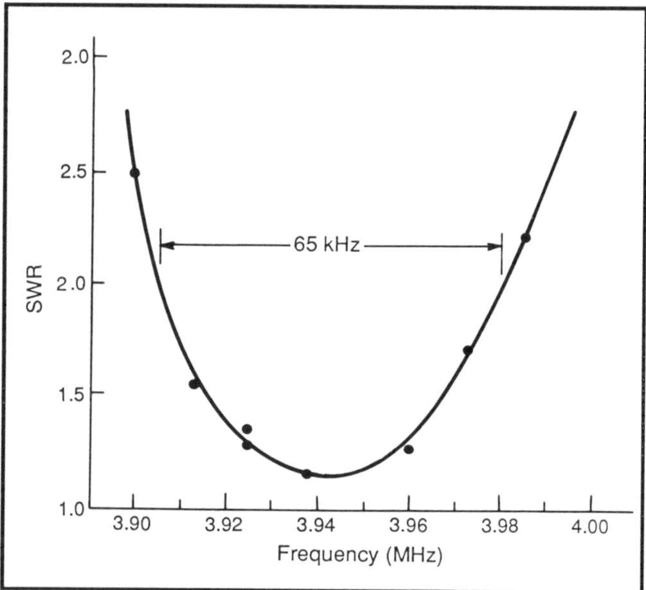

Figure II-7. The standing wave ratio for the 22-foot 80-meter vertical.

thought, I bought the umbrella and tried it out over my low-loss ground system. **Photo II-C** shows the final design shading my XYL; **Figure II-6** gives the parameters and VSWR curve. The antenna was matched to 50-ohm cable by a 16:1 unun. Although the bandwidth was somewhat less then that of the 6-footer, the on-the-air results were about the same. This antenna was quite a conversation piece!

As I stated earlier, all of the results obtained on 40 meters can be obtained on the other bands with proper scaling. I tried this out on 80 meters. Because a 7-foot top hat (rather than an 8-foot model) was available, the height actually turned out to be 22 feet instead of 20. The loading coil had 24 turns and was placed two feet below the top hat. The bandwidth was 65 kHz (half that of the 40-meter 10-footer) as is shown in **Figure II-7**.

Conclusions

Several interesting results came out of this investigation which, even to one schooled in antenna theory, are difficult to believe. I refer specifically to the ability of very short verticals to radiate and receive as well as a full-size quarter-wave antenna. The differences are practically negligible; but, the trade-offs are in lowered input impedances and bandwidths. However, with a good image plane and proper design, these trade-offs can be entirely acceptable.

Another point should be mentioned in relation to the results reported here. Even though my results were obtained using short ground-mounted verticals, they are valid for center-fed verticals or horizontal antennas, as well. The only difference is that one must double the impedance values and account for the effect of the image antenna.

I would like to acknowledge the support of many amateurs for their fine words of encouragement and excellent reporting during the on-the-air contacts. During the course of this study, some 350 amateurs reported on comparisons between signals from these short verticals and those of other stations.

References

1. King, *The Theory of Linear Antennas*, Harvard University Press, Cambridge, Massachusetts, 1956.
2. Weeks, *Antenna Engineering*, McGraw-Hill Book Company, 1968.
3. The diameter of the helical antenna was 1-5/8 inches. Each helical antenna had a top hat 18 inches above the helix in order to terminate the upper turns with sufficient capacitive reactance.

Appendix III

Short Ground-Radial Systems for Short Verticals

Appendix I reviewed my July 1971 *QST* article. The article, which investigated radials on the earth's surface, showed that four quarter-wave radials make a *very* poor ground system. With such a system, practically half of the output power is lost to the earth. It takes about 40 quarter-wave radials to reduce the loss to only a few ohms. Furthermore, the low-angle radiation is improved by the addition of more and longer radials. The article also showed that radials need not be buried to be effective. Because the electric field does not penetrate the earth's surface deeply at HF, the optimum radial depth is probably one which lets you cut the lawn safely, and also prevents the radials from becoming a tripping hazard to those using the yard. Additionally, it was noted that practically any size and type of conducting wire can be used to create a satisfactory ground-radial system.

Appendix II reviewed my March 1973 *QST* article, which dealt with the theory and practice of short, ground-mounted verticals. It investigated all the usual types of loading schemes and presented various short vertical antenna designs. Using measurements on verticals over an extensive ground system of 115, 0.4 wavelength radials, it revealed that a properly designed short vertical approaches the performance of a full-size quarter-wave vertical. The only real compromise is in bandwidth. Interestingly enough, my investigation also revealed that radials only 0.2-wavelengths long performed as well as the much larger 0.4-wavelength radials.

The major questions I had after writing these two articles were: 1) How short can the radials be and still yield acceptable results?, and 2) What is the soil conductivity under and in the near vicinity of the verticals I used? **Appendix III** reviews my April 1978 article, "Short Ground-Radial Systems of Short Verticals," which presents experimental evidence that answers these questions. My investigation showed that very small radial systems made up of almost any type of thin wire placed on the earth's surface can perform surprisingly well. On-the-air comparisons with a much larger vertical over an extensive ground system showed performances reduced by only a few decibels. Also, a new soil conductivity measurement technique, introduced for the first time in this article, indicated that the conductivity at my location is average to a little above average. Incidentally, several years after this article appeared in *QST*, I was informed that the soil conductivity technique presented in my article had been selected as the world standard.

Introduction

Vertical antennas have enjoyed considerable success on the 80- and 160-meter bands due to the difficulty of erecting horizontal antennas at heights sufficient for low-angle radiation. Optimum heights, which are in excess of one-half wavelength on these bands, are impractical for most amateurs. Often short verticals are used because they have been shown to compare favorably with full-size quarter-wave verticals. This is true if losses in the ground system, matching networks, and loading elements are small compared to the radiation resistance of the short vertical.

Considerable information is available that describes the effects of buried radials on the efficiency of quarter-wave verticals in the MF and LF bands as a function of the length and number of radials, and the conductivity of the soil.[1-17] However, there is little information on radials lying on the ground's surface—particularly in connection with short verticals.

Soil Conductivity

The conductivity of the soil under and in the near vicinity of a ground-mounted vertical is important in determining the extent of the radial system required

Material	Conductivity (millimhos/meter)
Poor soil	1–5
Average soil	10–15
Very good soil	100
Salt water	5000
Fresh water	10–15

Table III-1. General classification in conductivity.

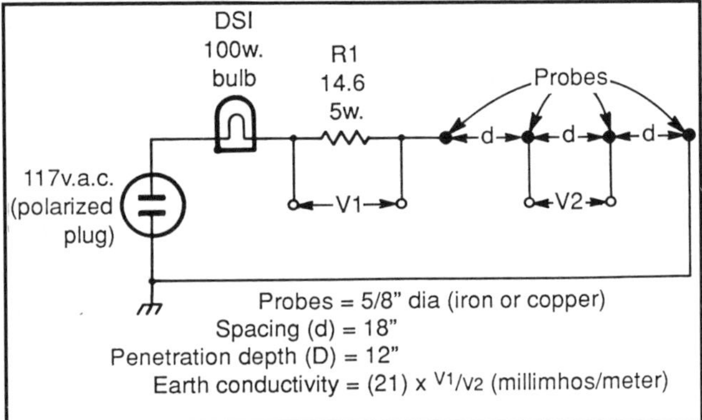

Figure III-1. Schematic diagram of four-point probe method for measuring earth conductivity.
DS1: 100-watt light bulb
R_1: 14.6 ohms (5 watts)
Probes: 5/8-inch diameter (iron or copper)
Spacing: d = 18 inches; penetration depth, D = 12 inches
Earth conductivity = 21 x V_1/V_2 (millimhos/meter).

and the overall performance that will be achieved. Most soils are nonconductors of electricity when completely dry. Conduction through the soil results from conduction through the water held in the soil. Thus, conduction is electrolytic. DC techniques for measuring conductivity are impractical because they tend to deplete the carriers of electricity in the vicinity of the electrodes. The main factors contributing to the conductivity of the soil are:

1) Type of soil.
2) Type of salts contained in the water.
3) Concentration of salts dissolved in the contained water.
4) Moisture content.
5) Grain size and distribution of material.
6) Temperature.
7) Packing density and pressure.

Although the type of soil is an important factor in determining its conductivity, rather large variations can take place from one location to another because of the other factors involved. Generally, loams and garden soils have the highest conductivities. These are followed in order by clays, sand, and gravel. Soils have been classified as shown in **Table III-1**. Although some differences are noted in the reporting[18,19] of this mode of classification because of the many variables involved, the classification generally follows the values shown in the table.

Because conduction through the soil is almost entirely electrolytic, AC measurement techniques are preferred. Many commercial instruments that use AC techniques are available and described in the literature.[20] However, there are rather simple AC measurement techniques, which provide accuracies on the order of 25 percent and are quite adequate for the radio amateur. Such a setup was developed by a Bell Labs colleague and neighbor, M.C. Waltz,[21] W2FNQ. It is shown schematically in **Figure III-1**. **Figure III-2** shows the conductivity readings taken over the last three months of 1976. It is interesting to note the general drop in conductivity over the three months studied, as well as the short-term changes due to periods of rain. The results presented in the following sections on antenna efficiencies were obtained during the period from October 10 to November 10, 1976, when the conductivity varied between 22 and 25 millimhos/meter. As **Table III-1** shows, my soil can be classified as average to slightly above average in conductivity.

Antenna Efficiency Considerations

The antenna efficiencies I will describe are based upon the losses that appear in series with the radiation resistance of resonant verticals. Although this approach does not give a comparison between the very low angles of radiation (i.e., less than 15 degrees) of various radial systems, it does allow for comparisons in the 15- to 30-degree range—which is important for sky-wave transmission on the 40-, 80-, and 160-meter bands. Mathematically, this definition for antenna efficiency can be written as:

$$\text{Antenna efficiency} = \frac{R_{rad}}{R_{rad} + R_g + R_A} \qquad \text{(Eq III-1)}$$

where:
R_{rad} = radiation resistance
R_g = ground loss
R_A = ohmic losses due to loading and antenna itself.

With high-Q loading coils and practically any size of aluminum tubing for the antenna, R_A can be minimized and therefore eliminated from the relationship shown above.

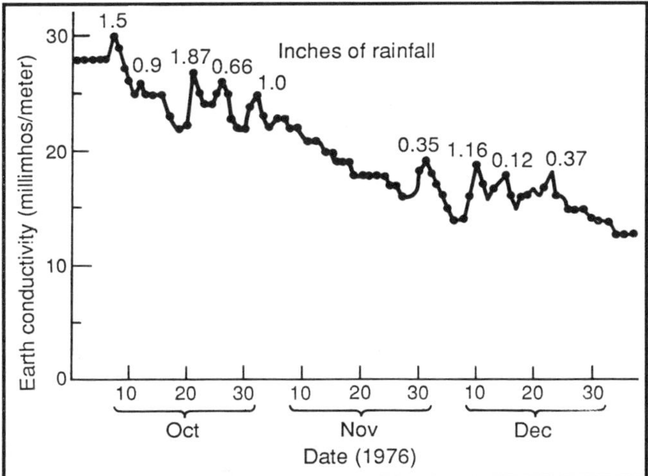

Figure III-2. Earth's conductivity at author's location during last three months in 1976.

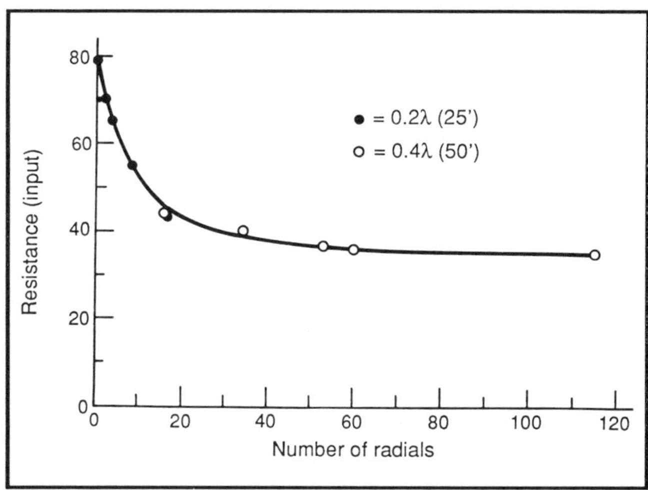

Figure III-3. Input impedance of resonant quarter-wave vertical as a function of the number of radials.

An example of this technique for determining antenna efficiency uses the results shown in **Figure III-3**. The input impedance of a resonant quarter-wave vertical is plotted as a function of the number of radials. Two lengths of radials (0.2- and 0.4-wavelengths) were considered. Because the radiation resistance for the thickness of the vertical used in this experiment is 35 ohms, the losses with 50 radials were approximately 2 ohms. With 100 radials, losses were about 1 ohm. This amounts to efficiencies of 94 and 97 percent, respectively. The efficiency with only 4 radials is less than 60 percent. This poor efficiency level exists even for a location with a soil conductivity that can be considered average to a little above average.

Furthermore, the efficiency of a radial system employing small numbers of radials is quite dependent on the moisture content of the soil. **Figure III-4** shows what happens with a resonant quarter-wave vertical on 20 meters, as the number of radials varies from one to eight. The difference in efficiency between wet and dry conditions becomes less pronounced as the number of radials is increased. The efficiency also becomes more independent of soil conductivity as the number of radials is increased.

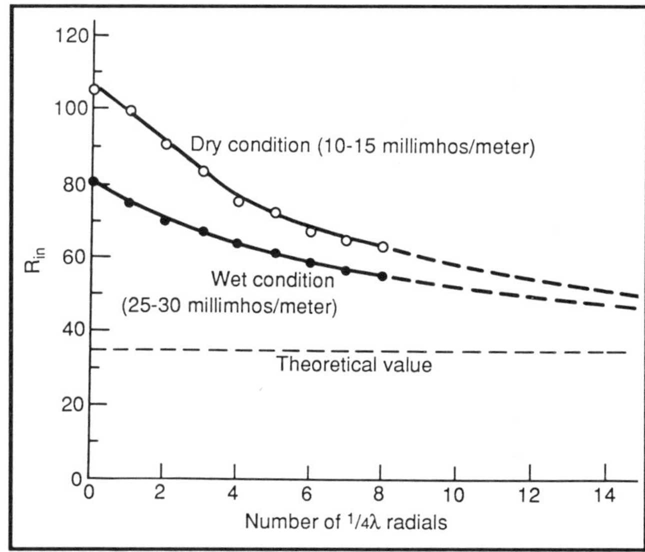

Figure III-4. Input impedance of resonant quarter-wave vertical as a function of the number of radials and the condition of the soil.

To determine the efficiencies of shortened verticals over abbreviated radial systems, I compared resonant input impedances with similar antennas over a near-ideal radial system. **Figure III-5** shows the experimental results for the various kinds of loading, as a function of height, over a near-ideal image plane (115 radials on the ground, about 50 feet long, and terminated in 10- to 12-inch nails). These results, which have been most useful to me in designing short verticals, are described in more detail in **Appendix II.**

Because all of my radials were on the surface of the earth and terminated in large nails, I undertook a study to determine the effect of the length of the spike. I measured the efficiency for resonant quarter-wave verticals with small numbers of shortened radials on the 20- and 40-meter bands. The radials were four and eight feet long (1/16th wavelength), respectively. **Figure III-6** provides the results for four different depths of termination. The figure shows that depths of 10 to 12 inches should be sufficient for the

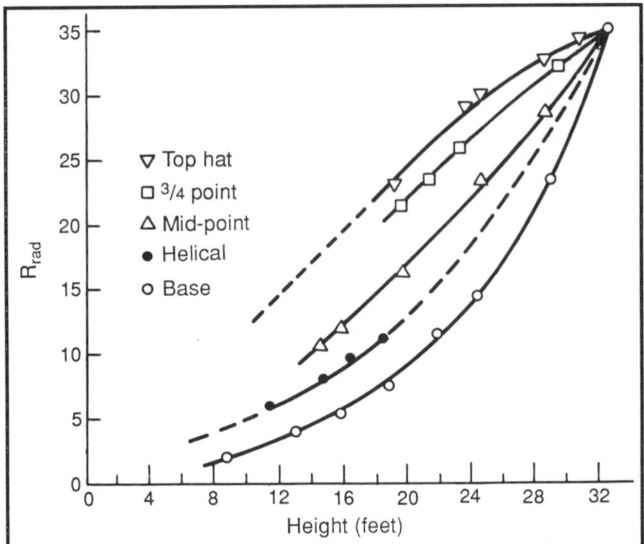

Figure III-5. Experimental results of radiation resistance as a function of height of antenna for various methods of loading.

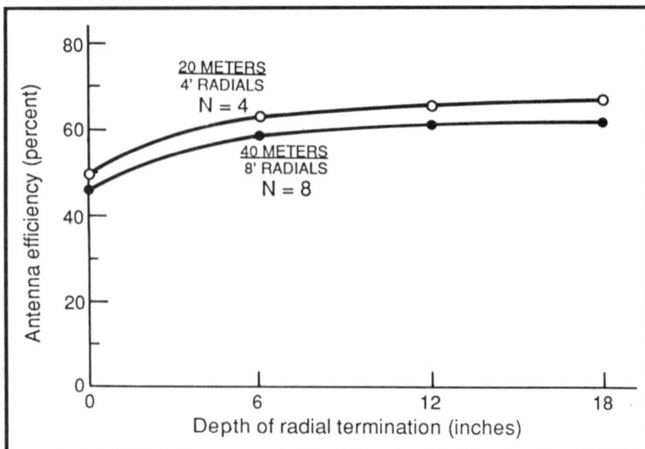

Figure III-6. Efficiency of resonant quarter-wave verticals on 20 and 40 meters as a function of the length of spike terminating the radials.

soil conductivity at my location for 20 and 40 meters, and most likely for 80 and 160 meters as well. Incidentally, the effectiveness of the radials of a quarter-wavelength or longer did not change noticeably as a function of the depth of the termination. Therefore, terminations for radials about 0.2-wavelengths and longer are primarily used for mechanical reasons.

Abbreviated Radial Systems

In order to determine experimentally the efficiency of shortened verticals with abbreviated radials on the earth's surface, I used five verticals of different heights and loading schemes on the 20- and 40-meter bands. My results were then compared with similar antennas over a near-ideal image plane as shown in **Figure III-5**. The five resonant verticals selected were:

1) 1/4 wavelength
2) 3/16 wavelength, top-hat loaded
3) 3/16 wavelength, midpoint loaded
4) 1/8 wavelength, top-hat loaded
5) 1/8 wavelength, midpoint loaded

Appendix II provides information on some of elements for these 10 verticals.

My test range used a radial system with various lengths of No. 17 steel fence wire terminated with 10- to 12-inch nails. **Photo III-A** shows the 12-inch aluminum base plate and the input connection arrangement. This antenna system was erected in the front yard approximately 50 feet from the house, and offered an opportunity for on-the-air comparisons

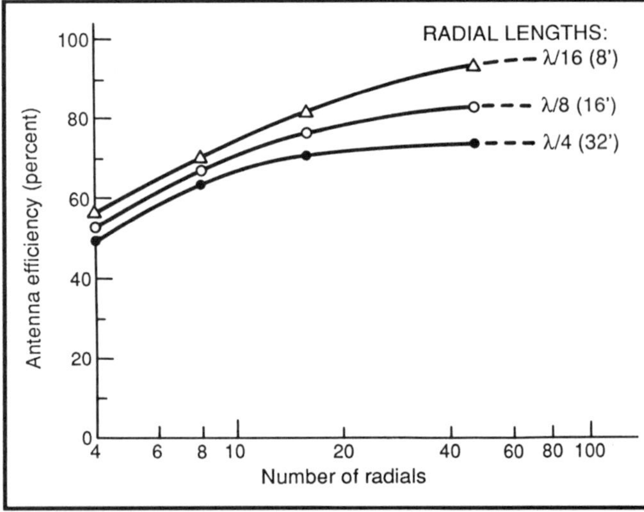

Figure III-7. Efficiency of resonant quarter-wave vertical as a function of the number of terminated radials with three different lengths.

with verticals mounted on the near-ideal ground system in the backyard.

The results are shown in **Figures III-7** and **III-8**. Only the 40-meter data is presented, as little difference was noted on 20 meters. **Figure III-7** shows the effect on the efficiency of a resonant quarter-wave vertical on 40 meters as a function of the number of radials using three different lengths of terminated radials. Although these curves were obtained on a 40-meter system, the relationships are generally valid for all other frequencies in the HF range if the same fractional wavelengths are used for the radials. As the figures show, the longer quarter-wave radials yielded the highest efficiency. However, it is interesting to note that this improvement in efficiency with length decreases as the number of radials decreases. At four radials, the efficiency with 8-foot radials is not much poorer than with 32-foot radials; i.e., 50 percent compared to 56 percent. Other interesting trade-offs also exist with various lengths and numbers of radials. **Figure III-7** shows that 16 quarter-wave radials are about equivalent to thirty-five 1/8-wave radials, and eight 1/4-wave radials are equivalent to only twelve 1/16-wave radials. Obviously, other equivalencies can be obtained from the figure.

Figure III-8 shows the results of antenna efficiency for short verticals as a function of the length of radials, with the number of radials kept constant at 48. As expected, 1/4-wave verticals, with their higher radiation resistance, have the highest efficiencies. Surprisingly, the efficiency of the 1/8-wave vertical does not suffer proportionally. That is, the 1/8-wave vertical with midpoint loading still has a 67 percent efficiency compared to the 84 percent efficiency of the 1/4-wave vertical, even though its radiation resistance is only one-third as large (12.5 compared to 35 ohms). A further comparison with a 1/4-wave vertical over an ideal ground system predicts that the 1/8-wave vertical with 48 1/8-wave terminated radials should show a reduced performance of only 1.7 dB.

Photo III-A. The 12-inch aluminum plate and input connection arrangement. Shown are 48 radials of No. 17 steel electric fence wire.

On-the-Air Comparisons

As was shown in the preceding section, short verticals with a sufficient number of abbreviated and terminated radials should yield performances of only a decibel or two poorer than quarter-wave verticals over near-ideal ground systems. Many on-the-air comparisons were made to confirm this prediction, which was based on efficiency considerations.

The first comparison involved a 40-meter, 1/8-wave, top-hat loaded vertical with forty-eight 1/8-wave (16 foot) radials (No. 17 steel wire on the ground and terminated with 10- to 12-inch nails). It was erected in the front yard. The input impedance of this vertical was 25 ohms and was matched with a highly efficient 2:1 unun. This antenna system was compared with one in the backyard using a 29-foot vertical with a 13.5-foot top hat over a ground system of 115, 50-foot terminated radials of No. 15 aluminum wire. This larger vertical was resonated by a small variable capacitor yielding an input impedance very close to 50 ohms. Over 100 contacts on 40 meters plus 200 observations on reception showed that the differences between these two systems were generally negligible. A few reports showed a 1- to 2-dB difference in favor of the much larger system, but these were in the minority.

An even more interesting comparison was made on 80 meters. A 20-foot vertical with an 8-foot top hat was erected in the front yard over the same ground system of 48, 16-foot, No. 17 steel wire radials (terminated). A base-loading coil of about 20 turns (6 tpi) of No. 12 wire with a diameter of 2.5-inches was required to resonate the antenna. The input impedance was close to 12 ohms. A 4:1 unun was used to match it to 50-ohm cable. This antenna system represents a radiation resistance of about 5 ohms and a ground loss of about 7 ohms. The high-Q loading coil had virtually no appreciable loss. This antenna was compared to the vertical in the backyard, which was 29 feet tall and had a 13.5-foot top hat and was over a near-ideal ground system. This taller antenna required about eight turns of No. 14 wire on a powdered-iron core (T200-2) to resonate on 80 meters. Its input imped-

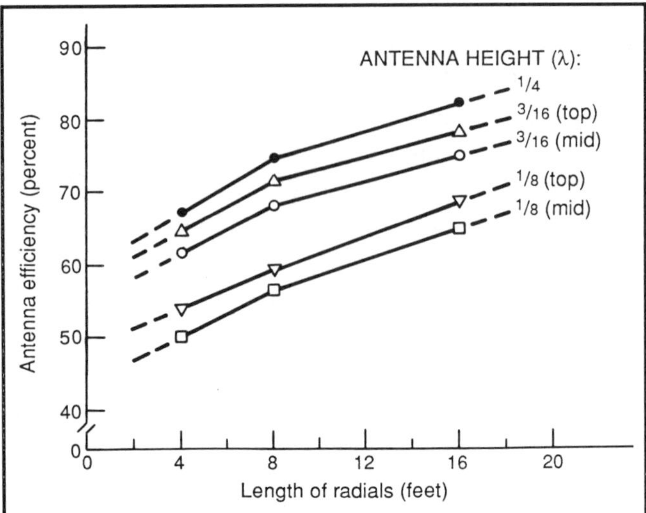

Figure III-8. Efficiency of short verticals as a function of the length of the terminated radials, with the number kept constant at 48.

ance was 15 ohms (showing negligible loss in the ground system and loading coil), and matching was accomplished with an efficient 3.33:1 fractional-ratio unun. Again, about 100 comparisons were made on the air and another 200 observations were made on reception. The average difference between the two systems amounted to only about 5 dB in favor of the much larger system in the backyard. This is quite noteworthy as many contacts with the larger system established it as a very competitive antenna system.

Concluding Remarks

Quarter-wave verticals over an extensive radial system have been known to be efficient, low-angle radiators. Even short verticals over the same large ground system have been shown to lose little in the way of performance. With low-loss ununs and matching techniques, short verticals suffer only in bandwidth. However, full-sized quarter-wave verticals and near-ideal ground systems are beyond the reach of most radio amateurs on 80 and 160 meters. This investigation was undertaken because little information was available on limited radial systems, particularly for short verticals.

As was shown, short radials over soil of average to a little above average conductivity can perform quite acceptably for verticals of all heights. The results of this investigation now let one predict quite accurately the operation of verticals less than a 1/4-wave and with radials as short as 1/16-wave. The simple soil-conductivity measurement technique described also provides a tool for comparing a given location with others, as well as predicting the performance of a ground-mounted vertical antenna.

Multiband vertical antennas, which have recently become commercially available, have demonstrated that the radial systems described here are not required because they operate electrically as 1/2-wave antennas. This is very likely true. Because it's nearly impossible to use this principle on 80- and 160-meters, the shortened vertical with abbreviated radials still appears to be the solution for low-angle radiation for many, if not most radio amateurs.

References

1. Brown, Lewis, and Epstein, "Ground Systems as a Factor in Antenna Efficiency," *Proc. IRE* 25, pages 753–787, June 1937.

2. Abbott, "Designs of Optimum Buried RF Ground Systems," *Proc. IRE*, pages 846–852, July 1952.

3. Wait and Pope, "Input Resistance of LF Unipole Aerials," *Wireless Engineer* 32, pages 131–138, May 1955.

4. Wait and Pope, "The Characterization of a Vertical Antenna With a Radial Conductor Ground System," *Appl. Sci. Research* (B)4, pages 177–195, 1954.

5. Monteath, "The Effect of the Ground Constants and of an Earth System on the Performance of a Vertical Medium-Wave Aerial," *Proc. IEE*, Vol. 105C, Pt 2, pages 292–306, January 1958.

6. Smith and Devaney, "Fields in Electrically Short Ground Systems: An Experimental Study," *J. Res. NBS* 63D Radio Prop. No. 2, pages 175–180, September/October 1959.

7. Larsen, "The E-Field and H-Field Losses Around Antennas With a Radial System," *J. Res. NBS* 66D Radio Prop. No. 2, pages 189–204, March/April 1962.

8. Maley and King, "Impedance of a Monopole Antenna With a Circular Conducting-Disk Ground System on the Surface of a Lossy Half Space," *J. Res. NBS* 65D Radio Prop. No. 2, March/April 1961.

9. Maley and King, "Impedance of a Monopole Antenna With a Radial-Wire Ground System on an Imperfectly Conducting Half-Space, Part I," *J. Res. NBS* 66D Radio Prop. No. 2, March/April 1952.

10. Wait and Walters, "Influence of a Sector Ground Screen on the Field of a Vertical Antenna," *NBS* Monograph 60, April 15, 1963.

11. Maley and King, "Impedance of a Monopole Antenna With a Radial-Wire Ground System on an Imperfectly Conducting Half-Space, Part II," *J. Res. NBS* USNC-URSI, 68D, No. 2, February 1964.

12. Maley and King, "Impedance of a Monopole Antenna With a Radial-Wire Ground System on an Imperfectly Conducting Half-Space, Part III, *J. Res NBS* USBC-URSI, 68D, No. 3, March 1964.

13. Gustafson, Chase, and Balli, "Ground System Effect on High Frequency Antenna Propagation," *Research Report*, U.S. Navy Electronics Laboratory, January 4, 1966.

14. Jager, "Effect of the Earth's Surface of Antenna Patterns in the Short Wave Range," *Int. Elek. Rundsch*, 24(4), pages 101–104, (1970).

15. Hill and Wait, "Calculated Pattern of a Vertical Antenna With a Finite Radial-Wire Ground System," *Radio Science*, Vol. 8, No. 1, pages 81–86, January 1973.

16. Rafuse and Ruze, "Low Angle Radiation from Vertically Polarized Antenna Over Radially Heterogeneous Flat Ground," *Radio Science*, Vol. 10, pages 1011-1018, December 1975.

17. Stanley, "Optimum Ground Systems for Vertical Antennas," *QST*, December 1976.

18. Card, "Earth Resistivity and Geological Structure," *Electrical Engineering*, pages 1153–1161, November 1935.

19. *Reference Data For Radio Engineers*, Fifth Edition, Howard W. Sams and Co., Inc., ITT, pages 26-3 to 26-5.

20. *Earth Resistances*, G. F. Lagg, Pitman Publishing Corp., 1964, pages 206–229.

21. Private communication.

Appendix IV

The Loading Coil

In 1989, after completing work on the second edition of my book *Transmission Line Transformers*, I thought it would be interesting to look at short vertical antennas for mobile use in the HF band and ground planes made from steel fencing and screening for field day use. I had purchased 20-, 40-, and 80-meter low-power (400 watt) Hustler whip antennas 10 years before, and had always intended to equip my car for mobile operation. Now, I finally had the opportunity to characterize these whip antennas over a large ground system of eight radials of steel fencing (with 2-inch x 3-inch openings) three feet wide and twenty five feet long, resulting in a ground loss of only 2 to 3 ohms on the three bands. Then I acquired the three Hustler high-power (2 kW) resonators to see what improvements they would make in the antennas' efficiencies. To my surprise, the differences in the input impedances of the 20- and 40-meter antennas were just about the same with either the low- or high-power resonators. This meant that the losses were comparable. However, there was a large difference in the losses between the 80-meter resonators. The 2-kW resonator had an input impedance of 35 ohms, while the 400-watt resonator had only 21.5 ohms. This indicated that the high-power loading coil could have about 15 ohms more loss! To better understand what the input impedance measurements meant, I decided to try to obtain a Q meter and study the fundamentals of the loading coil itself.

I borrowed a Boonton 260-A Q meter from a local amateur and began a characterization of the Hustler resonators and many other loading coils. Because the Boonton Q meter was an older (circa 1940) vacuum-tube model, accurate readings were difficult to obtain due to the necessary corrections needed in the Q and frequency readings. Luckily, my colleagues at AT&T Bell Laboratories had a later (circa 1970) solid-state model, a Hewlett-Packard 4342A Q meter, that I was able to borrow for a few weeks. This model didn't require me to make any corrections to the readings and, therefore, the readings were not only more accurate, but were also obtained in much less time.

Just as I was about to publish the results of my investigation, I received considerable correspondence containing critical feedback from readers regarding the second edition of my book. These concerns all required immediate attention. The readers were all having problems obtaining the components for my balun and unun designs. Most of them were not available retail! By working closely with Amidon Associates, I was able to redesign practically all of the transformers in my book with components that were readily accessible. In the process of redesigning many of the transformers, I came across new terms in the literature like *voltage*, *current*, and *choke* baluns—with which I was not familiar. After a more serious review of the amateur literature, I decided to not only publish my new designs, but my views on these new terms as well. The results of my new designs and views are the basis for this book.

Because my work on transmission line transformers is now complete (at least for the time being), I thought it might be interesting to present some of the highlights of my investigation on the loading coil. Obviously, what I've learned is relevant to the preceding three appendices.

I'll begin by discussing the theoretical considerations with loading coils and Q meters. I'll follow this with experimental results that include the Hustler resonators, B&W* coils, homemade coils and the effects of the mounting hardware. As you will see, the mod-

**Barker & Williamson, Inc., 10 Canal Street, Bristol, Pennsylvania 19007.*

eling technique I have used indicates that high-Q coils can reach values in the 800 to 900 range. These values are generally two to three times greater than expected. They are also two to three time greater than the values that have appeared in the literature.

Theoretical Considerations

Figure IV-1A shows the actual schematic of a coil and **Figure IV-1B** its equivalent series schematic. In **Figure IV-1A**, R_{ac} equals the AC coil resistance (which varies with frequency), L_o equals the low-frequency inductance (where C_d plays no role), and C_d equals the distributed capacitance (the capacitance between different parts of the coil).

Applying the usual circuit analysis to the parallel network of **Figure IV-1A**, we have for the impedance:

$$Z = \frac{R_{ac} + J\omega\,[L_o\,(1-\omega^2 L_o C_d) - C_d R_{ac}{}^2]}{(1-\omega^2 L_o C_d)^2 + \omega^2 C_d^2 R_{ac}{}^2} \qquad \text{(Eq IV-1)}$$

where:
$\omega = 2\pi f$

At frequencies that do not exceed 80 percent of the self-resonant frequency of the coil, **Equation IV-1** reduces to:

$$Z = \frac{R_{ac}}{(1-\gamma^2)^2} + \frac{J\omega L_o}{(1-\gamma^2)} \qquad \text{(Eq IV-2)}$$

where:

$$\gamma^2 = \left(\frac{f}{f_o}\right)^2 \qquad \text{(Eq IV-3)}$$

and

$$f_o = \text{self-resonant frequency} = \frac{1}{2\pi}\sqrt{\frac{1}{L_o C_d}} \qquad \text{(Eq IV-4)}$$

Thus, from the equivalent circuit of **Figure IV-1B**, we see that:

$$R_e = \frac{R_{ac}}{(1-\gamma^2)^2} \qquad \text{(Eq IV-5)}$$

and

$$L_e = \frac{L_o}{(1-\gamma^2)} \qquad \text{(Eq IV-6)}$$

and

$$\frac{\omega L_e}{R_e} = \text{equivalent Q of a coil} = Q_e \qquad \text{(Eq IV-7)}$$

The Q meter actually uses the equivalent circuit of **Figure IV-1B** in a series arrangement to bring about series resonance. The series equivalent circuit is also most easily visualized in loading-coil applications. **Figure IV-2** shows the series resonant circuit of a Q meter.

At resonance:

$$X_C = X_L \qquad \text{(Eq IV-8)}$$

and

$$i = e/R_e \qquad \text{(Eq IV-9)}$$

Therefore, the voltage across the variable capacitor at resonance becomes:

$$E = e/R_e \bullet X_C \qquad \text{(Eq IV-10)}$$

and

$$X_C/R_e = E/e = Q_e \qquad \text{(Eq IV-11)}$$

Equation IV-11 is correct for values of Q equal to or greater than 10. Therefore, if e is held to a constant and known level, a voltmeter can be connected across the capacitor and calibrated directly in terms of Q.

However, there is still one important problem with Q meters. They cannot measure coils with very high Qs. Hewlett-Packard[1] acknowledged this problem and said that they stop measuring coils when the Qs exceed 500. Therefore, to obtain some acceptable values for high-Q coils, I resorted to a curve-fitting technique for R_{ac}, which is the most difficult component to obtain. L_o and low-frequency values of R_{ac} are readily obtainable with the Q meter. The self-resonance and hence C_d are readily obtained with a gate-dipper and a modern transceiver.

The success in modeling the coil now rests upon the determination of R_{ac} with frequency. A good start is to assume that this parameter varies as the square-root of frequency as a result of *skin effect*.[2] Knowing accurate low-frequency values of R_{ac} and multiplying these values by the square-root of the frequency ratio yields:

$$R_{ac}\,|_{f_2} = R_{ac}\,|_{f_1}\left(\frac{f_2}{f_1}\right)^{1/2} \qquad \text{(Eq IV-12)}$$

where:
f_1 = the lower frequency where an accurate value of R_{ac} is obtained

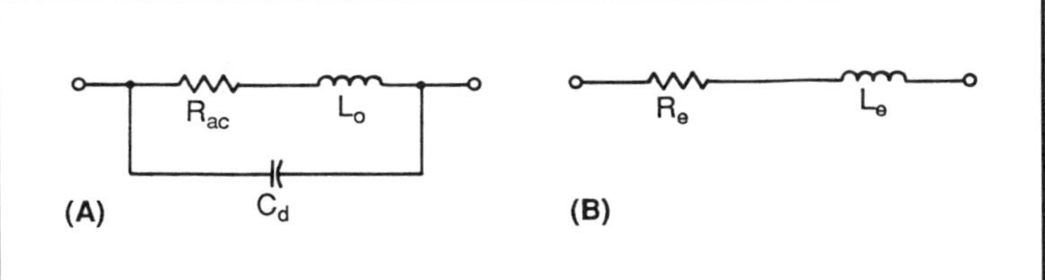

Figure IV-1. (A) is the actual schematic of a coil with distributed capacitance and (B) is the equivalent series schematic.

f_2 = the upper frequency where R_{ac} is calculated

By calculating several points within the region where accurate measurements are known to be achieved, one is able to determine if the calculated points are a good extension of the low frequency measurements. The square-root dependency works well for coils that have their turns spaced one wire diameter. For a tightly wound coil that has a considerable *proximity effect*[3] (crowding the current to the inner and outer surfaces of the conductors), the frequency dependency has been found to vary as the square of the cube-root of the frequency ratio, that is:

$$R_{ac} \mid f_2 = R_{ac} \mid f_1 \left(\frac{f_2}{f_1}\right)^{2/3} \qquad \text{(Eq IV-13)}$$

Thus, with hand calculators that have the function y^x, one is able to determine quite precisely what fractional power to raise the frequency ratio for a better fit to the curves.

Experimental Results

The first measurements and comparisons were made on the 80-meter loading coils, because I wanted to find out why the Hustler high-power whip antenna had a much higher input impedance over a low-loss ground system. The answer lies in **Figure IV-3**. At 3.5 MHz, the high-power resonator (loading coil) displayed a loss about twice that of the low-power unit—40 ohms compared to 20 ohms. Even though the RM-80-S resonator used a tightly wound, No. 14 Litz wire coil, its main loss resulted from using metal end-caps very close to the ends of the coil (less than 1/4 inch). As such, they not only induced excessive loss, but also significantly reduced the inductance of the coil as well. Even though the 400-watt unit used No. 20 wire wound tightly on a smaller coil form, its metal end-caps were considerably further from the ends of the coil, thus accounting for the higher efficiency.

Curve C shows what can be done with No. 14 H Thermaleze wound tightly on a 2-inch diameter plastic form (from the grocery store) with no end-cap effect. Finally, curve D shows what can be achieved with a high-Q coil that has a spacing of about one wire-diameter between the turns. **Figure IV-4** shows the variation in inductance and Q with frequency for the B&W, 80-meter loading coil. Even though the inductance was less than that of the other three (the

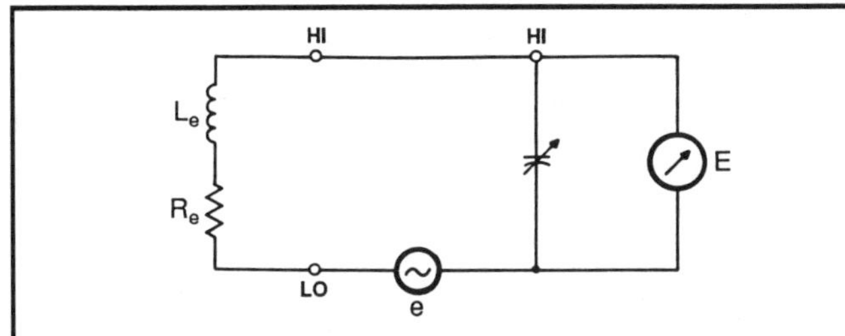

Figure IV-2. The Q meter series resonant schematic.

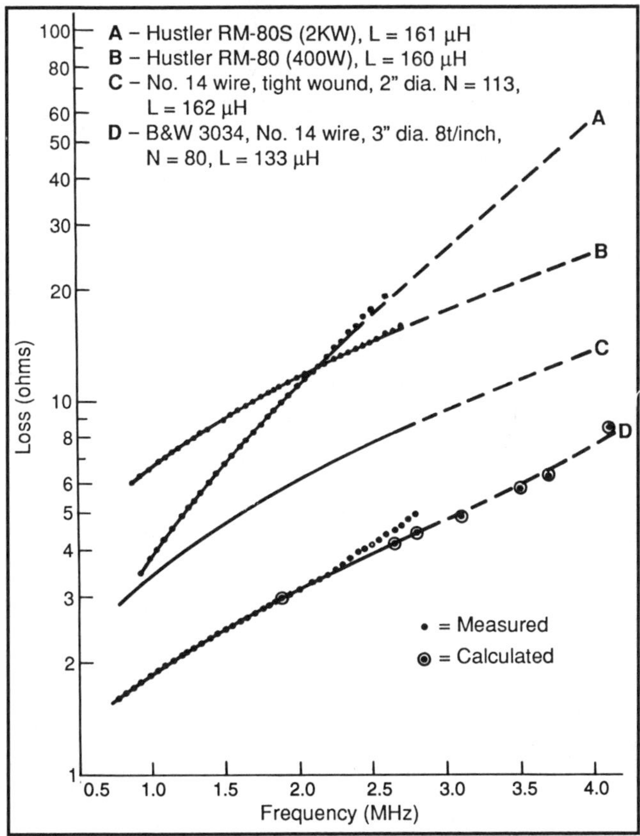

Figure IV-3. A comparison of losses with four different 80-meter loading coils used with whip antennas.

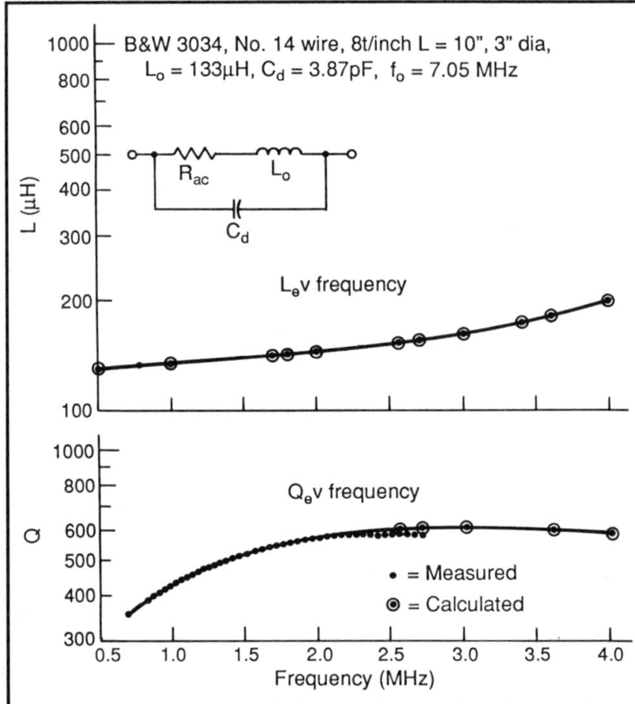

Figure IV-4. The variations in the equivalent inductance, L_e, and Q_e with frequency of the low-loss loading coil, B&W 3034, used in the 80-meter whip in Figure IV-3.

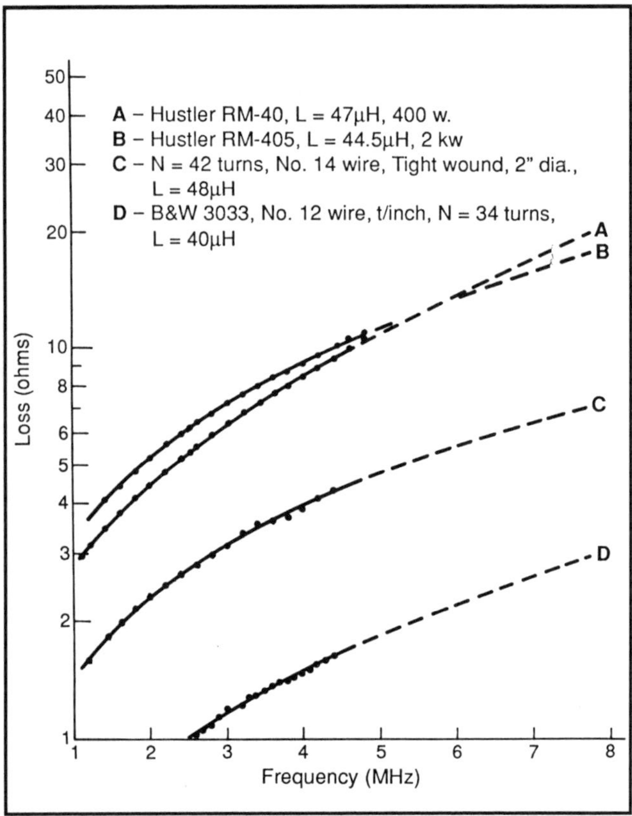

Figure IV-5. A comparison of losses with four different 40-meter loading coils used with whip antennas.

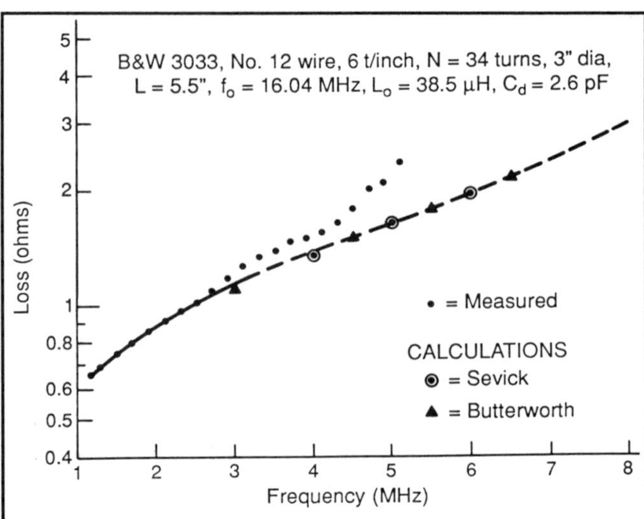

Figure IV-6. A later look at the variation in loss with frequency of the low-loss loading coil, B&W 3033, used in the 40-meter whip shown in Figure IV-5.

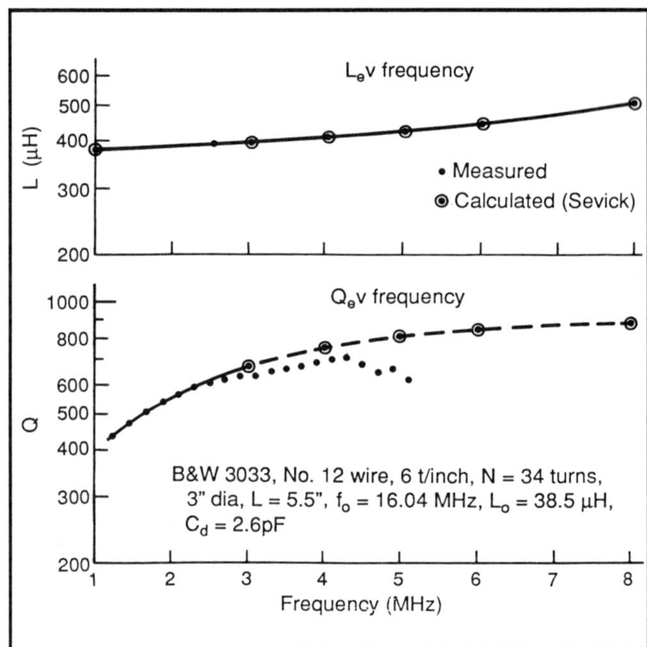

Figure IV-7. The variations in equivalent inductance and Q with frequency for the low-loss coil, B&W 3033, shown in Figure IV-6.

coil is only obtainable in 10-inch lengths), and therefore requires a little extra length above the coil for resonance, its improved efficiency is obvious.

Figure IV-5 shows some early comparisons with 40-meter loading coils. In this case, the low-power and high-power resonators show about the same loss at 7 MHz. Again, curve C shows what can be achieved by eliminating the end-cap effect. Curve D

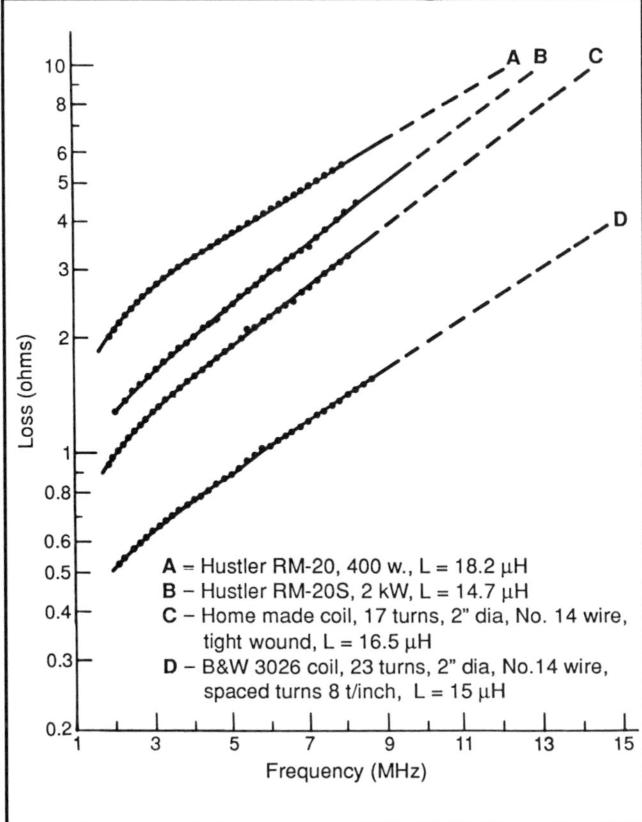

Figure IV-8. A comparison of losses with four different 20-meter loading coils used with whip antennas.

shows what happens when the turns are spaced. **Figures IV-6** and **IV-7** provide a later view of the B&W 40-meter loading coil. They also include some computations from Butterworth's work.[3] Note the disparity between the experimental and calculated results beyond 3 MHz where the Q is greater than 600!

Figure IV-8 shows early comparisons with the 20-meter loading coils. The high-power Hustler units use No. 14 spaced wire on a 2-inch diameter form. The low-power unit has many No. 20 close-wound turns on a 1/2-inch diameter coil form. Curve C shows what can be done by eliminating the end-cap effect and curve D, and also the effect of spacing the turns. **Figures IV-9** and **IV-10** show what can be obtained with a coil using a slightly thicker wire and larger diameter.

Figures IV-11 and **IV-12** show what can be obtained with 40-meter loading coils for mobile applications using No.16 wire. Curve C in **Figure IV-12**, the B&W 3035 coil, shows a Q approaching 700 at 8 MHz!

Figure IV-13 shows the loss in a loading coil as a function of the spacing of the hardware in a concentric mounting. The loss nearly triples when the spacing goes from zero to infinity (only the nylon rod

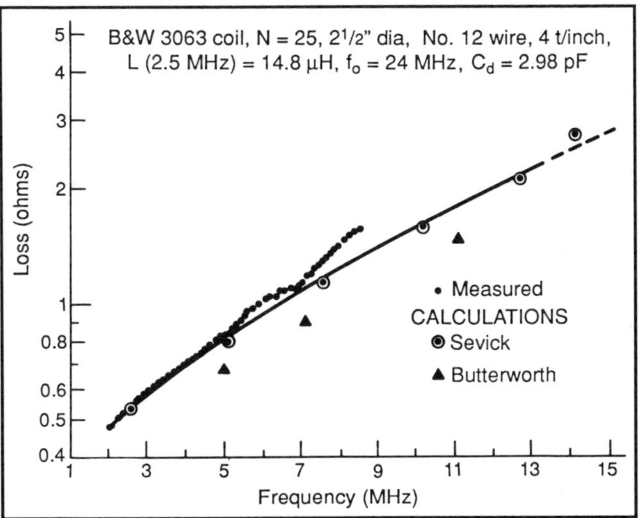

Figure IV-9. A later look at the variation in loss with frequency of another low-loss loading coil, the B&W 3063, used with the 20-meter whip. It has a little thicker wire (No. 12) and larger diameter than one shown in Figure IV-8.

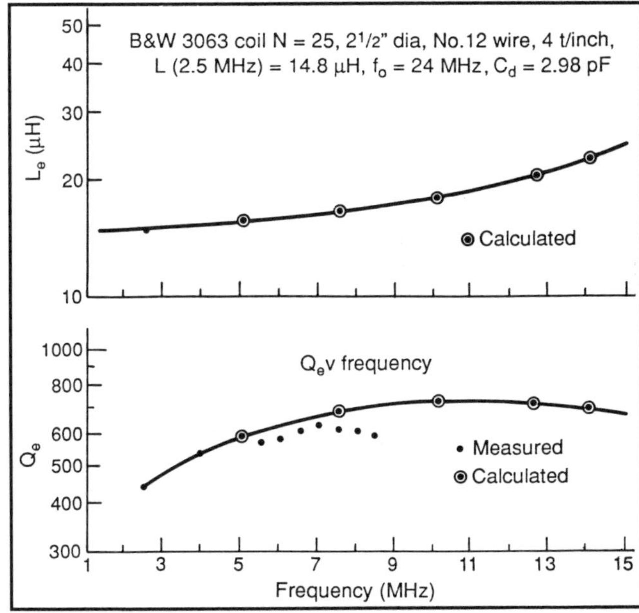

Figure IV-10. The variations in equivalent inductance and Q with frequency for the low-loss coil, B&W 3063, shown in Figure IV-9.

itself). **Figure IV-14** shows the loss versus frequency when a coil similar to the one in **Figure IV-13** is mounted 1.5-inches from the aluminum tubing and insulator (side-mounted). In this situation, the loss increases from 2.3 to 2.8 ohms at 3.5 MHz.

Figure IV-15 shows the experimental results of the sensitivity of loss to the spacing of metal end-caps. This experiment was prompted by the large loss

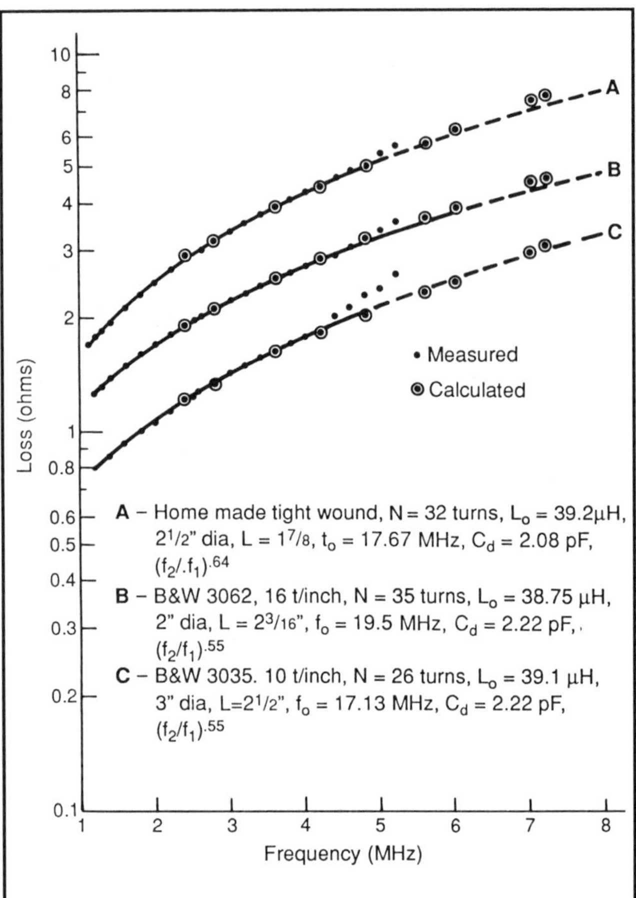

Figure IV-11. A comparison in the losses versus frequency of three loading coils using No. 16 wire with different spacings between turns (primarily).

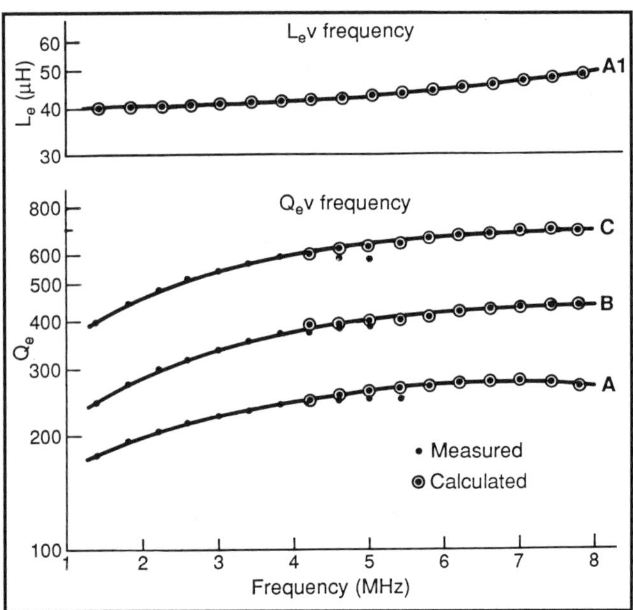

Figure IV-12. The variation in the equivalent Q with frequency of the three coils of Figure IV-11 and the variation in inductance with frequency for the tightly wound coil.

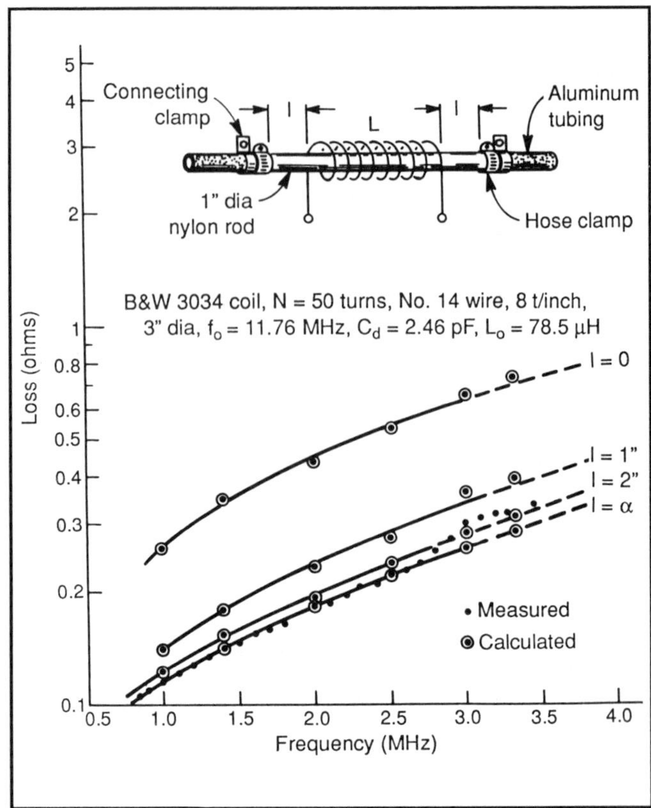

Figure IV-13. The variation in loss with frequency for different spacings of mounting hardware (the concentric case).

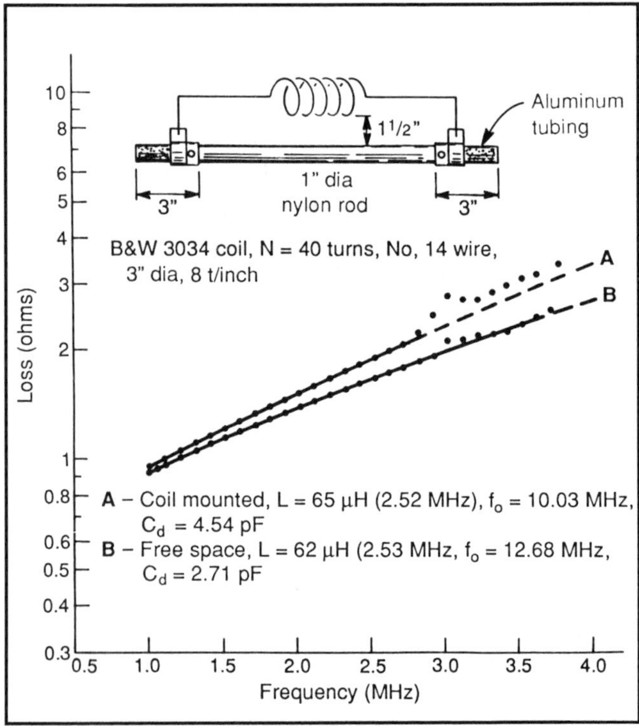

Figure IV-14. A comparison in the loss versus frequency between a coil mounted 1.5-inches from the insulator and aluminum tubing, and one in free space.

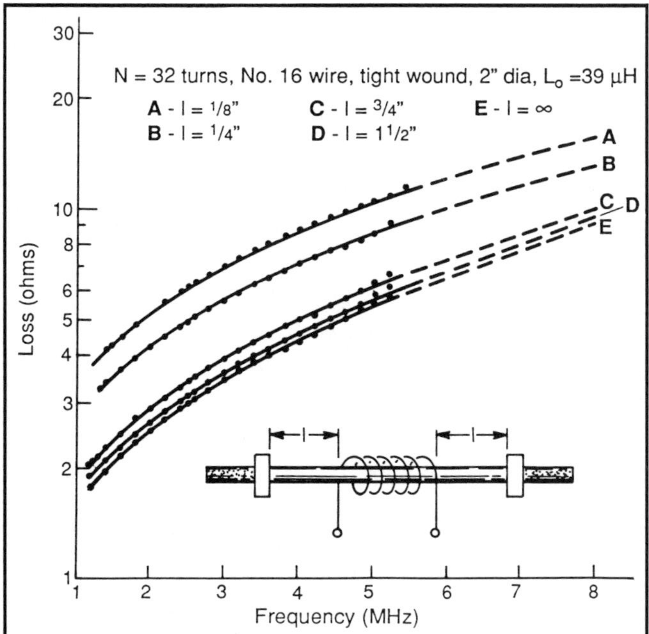

Figure IV-15. The variation in loss versus frequency for different spacings of aluminum end-caps.

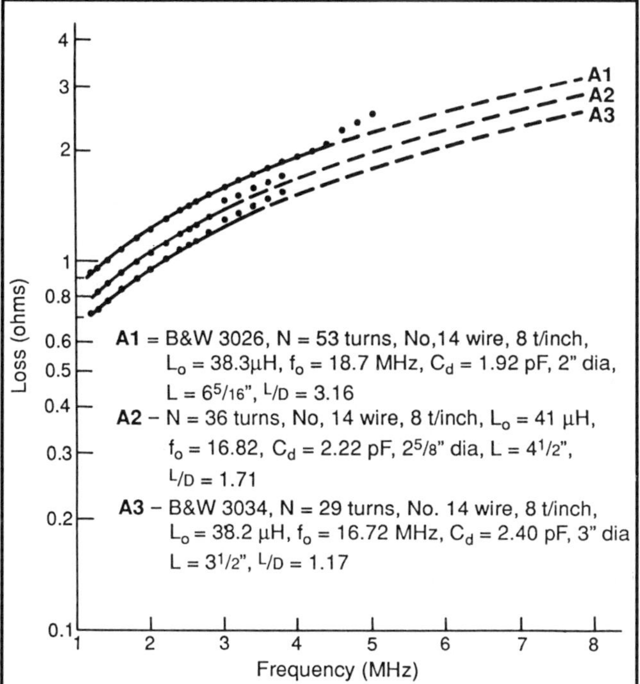

Figure IV-16. A comparison of the loss as a function of frequency and the length-to-diameter ratio, l/D.

exhibited by the Hustler high-power 80-meter resonator. At 3.5 MHz, the loss doubles in going from zero spacing to infinity (no end-caps at all). Although this information is not shown, I also investigated copper and steel end-caps. The results with copper were about the same as that of aluminum. However, steel

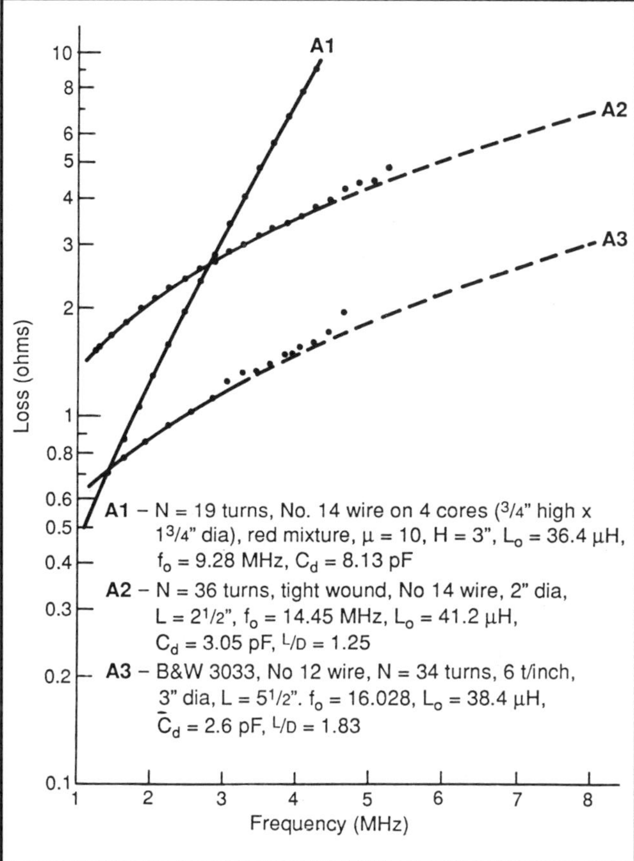

Figure IV-17. A comparison of loss versus frequency between a coil using a powered-iron core and two that are air-wound.

showed about three times as much loss as the other two metals.

Figure IV-16 attempts to show coil loss as a function of the length-to-diameter ratio, l/D, of coils with relatively the same inductance. Because considerable space has been devoted to this subject in the literature, I was surprised at the small difference observed in going from an l/D of 3.16 to 1.17. At 3.5 MHz, it shows an improvement of only 0.5 ohms. Even though the ideal l/D ratio is stated to be around 0.5, I expected larger differences then were seen in this investigation.

Another question came to mind regarding the use of powdered-iron cores for more compact inductors. Experiments with transmission line transformers showed that the red mixture, which has a permeability of 10, yields the same very high efficiency as a ferrite with a permeability of 40 (which yields the highest efficiency of practically all ferrites). **Figure IV-17** shows the results of comparisons between a coil using the red mixture and "air-wound" coils using tight and spaced windings. They all have about the same induc-

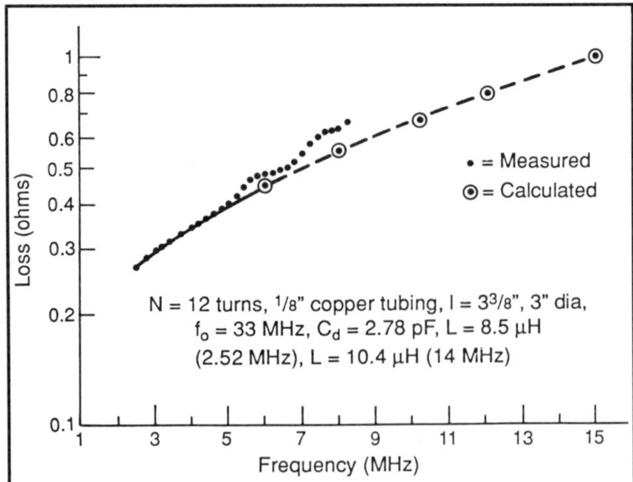

Figure IV-18. The variation in loss with frequency for a low-loss coil using 1/8-inch copper tubing.

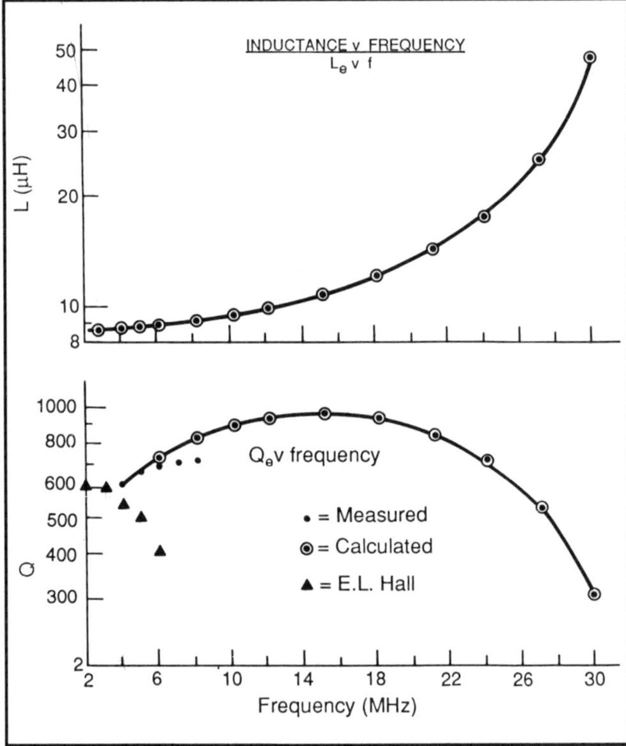

Figure IV-19. The variation in equivalent inductance and Q with frequency of the low-loss coil of Figure IV-18 using copper tubing.

tance. Even though the loss is excessive beyond 3 MHz, this experiment justifies its use on the 160-meter band.

An example of a very high-Q coil using 1/8-inch copper tubing is shown in **Figures IV-18** and **IV-19**. As **Figure IV-19** shows, the calculated equivalent Q exceeds 900 beyond 10 MHz. Note that the Hewlett-Packard Q meter becomes unreliable above 5 MHz where the Q exceeds 600. Also shown is some very early data from E.L. Hall on a similar coil which was published in Terman's popular handbook.[4]

A comparison of losses was also made on rods of different insulating materials. The rods used had 1-inch diameters and sufficient lengths so the coils could be mounted concentrically without adding any loss due to the connecting hardware. The coil was a B&W 3033 of 15 spaced turns of No. 12 wire. A nylon rod and an acrylic rod showed no extra loss above the loss that the coil had when it did not contain a rod. One-eighth-wall PVC and hollow phenol fiber exhibited approximately the same additional loss. The increase was about 1 percent at 5 MHz and about 2 percent at 10 MHz. Surprisingly, a solid phenol-fiber rod had a 10-percent increase in loss at 5 MHz and a 23-percent increase at 10 MHz. A dry wooden dowel and a 1/8-wall cardboard tube had approximately the same increases in loss—4.5 percent at 5 MHz and 10 percent at 10 MHz. However the wooden dowel, when wet, was something else. The increase in loss at 5 MHz was 92 percent, and the increase at 10 MHz was 340 percent!

Closing Comments

Just as my curiosity concerning the losses in Hustler whip antennas started me on a rather long investigation of the ubiquitous loading coil, it is only appropriate to end this appendix with a perspective on mobile whip antennas. **Table IV-1** shows how the Hustler antennas compare in efficiency with whip antennas using low-loss loading coils, mounted on the back-end of a car, which is parked on a macadam driveway. The resonant input impedance, R_{in}, is measured when the antennas are mounted over a low-loss ground system and when they are mounted on the back-end of a car. The radiation resistances, R_{rad}, were obtained from *The ARRL Antenna Book*[5] for center-loaded, 8-foot whip antennas. Although the 20-meter antennas were a little less than 7 feet in height, they were loaded at the 2/3rds point and therefore had a radiation resistance close to that of a center-loaded, 8-footer. On 20 meters, even the hustler whips are only poorer by about 5 dB from the ideal quarter-wave condition. When looking at the other numbers, the question comes up as to whether it's worth replacing the Hustler resonators with high-Q coils. There would only be a 3 dB improvement on 40 meters and 80 meters over the low-power unit. Even the lossy RM-

Antenna	R_{in} (over low-loss ground)	R_{in} (on car)	R_{rad}	eff.(%)
RM-80	21.5	23	0.8	3.5
RM-80S	35	38	0.8	2.2
B&W 3034	8	12	0.8	6.7
RM-40	21.5	26.5	3	11.3
RM-40S	20.5	25.5	3	11.8
B&W 3034	8	14	3	21.5
RM-20	18	31	11	35.5
RM-20S	19.5	32	11	34.4
B&W 3063	15	27	11	41.0

Table IV-1. Efficiency comparisons of mobile whip antennas.

80-S high-power unit is only down 5 dB from an efficient radiator of the same dimensions. Another surprise was the small increase in loss to ground in going from a low-loss ground system to the ground offered by a car.

Several other results stood out in this investigation. One was the sensitivity of the loss in a coil to the spacing between turns. Another was the sensitivity of the loss to the spacing of the mounting hardware. Because the length-to-diameter ratio, l/D, has appeared so much in the literature, it was a surprise to see the experimental results that showed it is not such a sensitive parameter. Finally, the Qs of low-loss coils are much higher than I expected. I don't think values of 800 to over 900 have appeared in the literature before. It will be interesting to see if this appendix prompts some responses from the readers.

References

1. Private communication.

2. Wheeler, "Formulas for the Skin Effect," *Pro. of the IRE*, September 1942, pages 412–424.

3. Butterworth, "The High Frequency Resistance of Toroidal Coils," *Wireless and Wireless Engineer*, Vol. 6, January 1929, page 13.

4. Terman, *Radio Engineers' Handbook*, McGraw-Hill Book Company, 1943, page 75.

5. Hall, *The ARRL Antenna Book*, 16th edition, The American Radio Relay League, Newington, Connecticut, page 16–4.

Index